T0211421

Nature-Inspired Optimization Algorithms

Nature-Inspired Optimization Algorithms

A Vasuki

CRC Press
Taylor & Francis Group
Boca Raton London New York

CRC Press is an imprint of the
Taylor & Francis Group, an **informa** business

A CHAPMAN & HALL BOOK

First edition published 2020
by CRC Press
6000 Broken Sound Parkway NW, Suite 300, Boca Raton, FL 33487-2742

and by CRC Press
2 Park Square, Milton Park, Abingdon, Oxon, OX14 4RN

ISBN: 978-0-367-25598-5 (hbk)
ISBN: 978-0-367-50329-1 (pbk)
ISBN: 978-0-429-28907-1 (ebk)

Contents

Preface

This book is intended as a reference for undergraduate and postgraduate students as well as researchers who are working on complex problems to find optimum solutions. The book can serve as a reference for faculty who are handling classes on the topic of optimization. It gives a general overview of optimization and its applications with simple examples and discusses conventional optimization techniques and their limitations. Fifteen nature-inspired optimization algorithms have been included with a detailed explanation of the techniques.

These optimization algorithms can be applied to any real-life problem, either constrained or unconstrained. It involves choosing values of parameters associated with the problem in order to arrive at the optimum solution. The unconventional optimization algorithms are broadly based on biological evolution and physical and chemical processes. These nature-inspired algorithms have been developed from the study of swarm behavior of animals, birds, and insects. The main advantage is that they are population-based, and hence the search for the optimum solution can be conducted in parallel by multiple agents. This reduces the time taken to arrive at the (best) global optimum solution to the problem. Since the problems could be multimodal with several local optima, the algorithm should be able to distinguish between local and global optimum solutions.

These algorithms can be tested with standard benchmark data sets and classical engineering design problems. The standard problems in computer science such as the traveling salesman, graph coloring, finding the shortest path in a graph, job scheduling, and routing in computer networks are NP-hard. Image processing is another broad area that includes enhancement, segmentation, compression, classification, object recognition, feature selection, clustering, and registration. The application of nature-inspired algorithms to such intractable problems leads to optimum solutions in the shortest possible time. A list of comprehensive problems upon which these algorithms can be tested is also provided.

I sincerely thank everyone who have helped me in completing this book.

Dr. A Vasuki

Author

Dr. A Vasuki is currently working as Professor in the Department of Mechatronics Engineering at Kumaraguru College of Technology, Coimbatore, India. She has 28 years of teaching, research and academic administration experience. She has completed B.E in Electronics and Communication Engineering from PSG College of Technology in 1989. She has completed her postgraduate degree M.E Applied Electronics from Coimbatore Institute of Technology in 1991. She has done her Ph.D in Image Compression from PSG College of Technology under Anna University Chennai in 2010. Her research interests are Signal Processing, Image Processing, Communication and Optimization. She has published 3 Book Chapters, 38 National and International Journal papers and 60 National and International Conference papers. She has guided 30 PG projects and 50 UG projects. She is an approved Research Supervisor under Anna University Chennai and is currently guiding 9 research scholars.

1

Introduction

1.1 Introduction

Optimization is of prime importance in most of our everyday activities and plays a vital role in achieving desired goals. Optimization is arriving at the best set of parameters, maybe by trial and error, so that we achieve the desired objective(s) exactly or at least a good approximation. We optimize parameters to achieve desired objectives without being consciously aware of it many times in our day-to-day activities. Optimization can be applied in diverse fields such as mathematics, physics, chemistry, engineering design, production, telecommunication, networks, computer science, economics, management, etc. The problems range from the very simple to the highly complex ones like designing steel structures, aircraft routes, road connectivity, routing in computer networks, unmanned space missions, autonomous vehicles, manufacturing, automation, robots, and so on. The goal of optimization could be to maximize quality/profit/efficiency or minimize cost/wastage/resources utilized with reduced time, space, and computational complexity. The problems might have constraints that have to be satisfied in finding the optimal solution. The constraints could be limits on length, area, power consumed, cost, weight, or restriction on certain variables to assume integer values (positive or negative), and some variables need to be bounded between minimum and maximum values as specified in the problem. In all optimization problems there will be several parameters that have to be adjusted to get the best possible output. The problems to be solved are generally complex in nature with many conflicting requirements. The solution attained has to accommodate all these various conflicting requirements, and, keeping the objective(s) in focus, the optimization technique has to be designed.

1.2 Fundamentals of Optimization

Optimization is obtaining the optimum values of a set of variables upon which an objective function depends, and that could be either constrained or unconstrained. Any set of values assigned to the variables always produces an output but the optimum set of values produces the optimum output. The objectives of the problem and the constraints, if any, can be formulated in terms of mathematical functions or equations. The mathematical expression representing the goal of optimization is called the objective function. The objective function could be defined for maximization or minimization depending on the problem domain. Figure 1.1 illustrates the general optimization problem with input

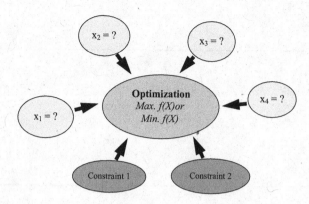

FIGURE 1.1
General optimization problem.

variables and constraints that either maximizes or minimizes the objective function. In Figure 1.1 the objective function $f(X)$ depends on the vector of input variables represented by $X = [x_1\ x_2\ x_3\ x_4]$.

When an objective function is to be maximized, it is called a fitness or quality function. When it is to be minimized, it is called a cost or penalty function. For a minimization function, if the minimum value is zero it is called an error function. The sign of the objective function can be complemented (+/−) in order to transform a maximization problem to minimization and *vice versa*. An objective function that is to be maximized can also be minimized by taking the negative of the function.

Some of the characteristics of the objective function are:

- The number of variables in the function is the dimension (d) of the search space.
- Whether the variables assume continuous or discrete values determines whether the function is continuous or discrete.
- If the function is continuous, is it differentiable at all points in the search space?
- Whether the function has one maxima (unimodal) or multiple maxima (multimodal).
- Is the function unconstrained or constrained? If constrained, how many and what are the constraints? Are they equality or inequality constraints?

Mathematically, a function is represented as $f: R^d \rightarrow R$, where the function f belongs to the d-dimensional hyperspace R^d. The domain R^d is the parameter or search space with each $X \in R^d$ being a possible candidate solution to the objective function $f(X)$. The function $f(X)$ maps the search space to the function space R. The problem is to find $X^* \in R^d$, for which $f(X^*) \geq f(X),\ \forall X \in R^d$. This applies to a maximization problem, and for minimization problems, it has to be $f(X^*) \leq f(X),\ \forall X \in R^d$. The search space could be either the entire d-dimensional hyperspace or a subset of the hyperspace. When there are constraints associated with the problem, the search space becomes a subset of the d-dimensional hyperspace. The characteristics associated with the constraints of the problem are:

- Number of constraints for the problem
- Whether the constraints are equality or inequality constraints

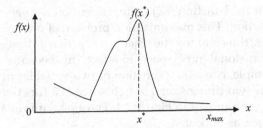

FIGURE 1.2
One-dimensional function $f(X)$ where $X = [x]$.

Figure 1.2 shows an objective function $f(X)$ where the vector $X = [x]$ is one-dimensional, since the function depends on only one variable x.

The one-dimensional plot of the objective function $f(X)$ shown in the Figure 1.2 has a maximum value at $x = x^*$. This is the global optimum since this is the highest value for the function $f(X)$. The search space is the one-dimensional x-axis where the continuous variable x ranges from *zero* to a maximum value x_{max}. In general, the search space is $S \rightarrow R^d$. The convergence of the algorithm or the final value attained is the optimum solution and depends to a great extent on the starting point in the search space or the initial value of x. If the negative of the function $f(X)$ is taken, the maximum value becomes the minimum value or the peak becomes a valley. The maximization problem becomes a minimization problem when the global minimum occurs at $X = x^*$.

Extending this concept further, the one-dimensional function $f(X)$ could have multiple maxima or minima, as illustrated in Figure 1.3. The function $f(X)$ shown in Figure 1.3 has multiple peaks and valleys, where $X = [x]$. The peak with the highest value is called the global maximum and the valley with the lowest value is called the global minimum. The other peaks which are smaller than the maximum are called local maxima, and the valleys which have higher values than the minimum are called local minima. The search for the optimum solution for $f(X)$ starts somewhere along the curve and the objective is to find the global maximum or minimum, as the case may be. The problem occurs when the search for the optimum solution gets trapped in local maxima or minima. The optimization algorithm should be such that the function should attain the global optimum value and if it gets trapped in local optimum it should be able to come out of the local optimum point or region in the search space.

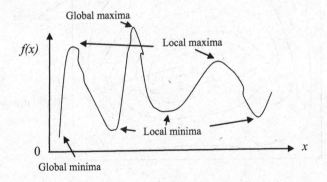

FIGURE 1.3
One-dimensional function $f(X)$ having multiple maxima and minima.

In the above problem the function $f(X)$ is dependent on a single variable x; hence it is single variable optimization. This maximization problem of one variable can be extended to d number of variables, thus making the objective function d-dimensional and the search space becomes a d-dimensional hyperspace. In effect, this becomes multivariable optimization. As another example, consider a function of two variables $f(X) = f(x_1, x_2)$ with only one global minima. The two-dimensional search space of a function of two variables $f(X) = f(x_1, x_2)$ is shown as a contour plot in Figure 1.4. The global minima appears nearly at the center of the search space, as indicated.

In general,

$$f(X) = f(x_1, x_2, \ldots x_d) \tag{1.1}$$

and the search space is $S \rightarrow R^d$. This is single-objective optimization problem since there is only one objective function that is to be maximized or minimized. In contrast to this, if there is more than one objective function to be maximized or minimized then it becomes multi-objective optimization. If there are no constraints attached to the problem it is unconstrained optimization whereas if there are one or more constraints in the problem it is constrained optimization. The constraints of the problem could be equality or inequality constraints. They are mathematically represented as:

$$g_i(X) = 0, \quad i = 1, 2, \ldots P$$

$$h_j(X) \geq 0, \quad j = 1, 2, \ldots Q \tag{1.2}$$

$$or \ h_j(X) \leq 0, \quad j = 1, 2, \ldots Q$$

where the number of equality constraints is P and the number of inequality constraints is Q. When there are multiple number of objective functions, the problem is multi-objective optimization. This is mathematically represented as:

$$f(X_k) = f(x_{k1}, x_{k2}, \ldots, x_{kd}), \quad k = 1, 2, \ldots K \tag{1.3}$$

where K is the number of objective functions that are to be either maximized or minimized. The objective function $f(X)$ maps the search space to the function space, which will

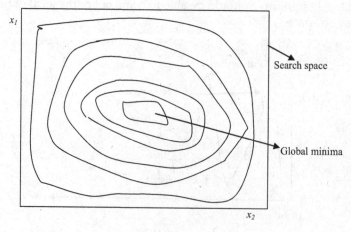

FIGURE 1.4
Two-dimensional search space of a function $f(x_1, x_2)$ with one global minimum.

have a single value for single-objective optimization and multiple values in the case of multi-objective optimization. The parameters or variables can be continuous or discrete and belong to a finite or infinite set. The parameters could be interdependent or they may be independent of each other. The function might have only one maximum point that is the global maximum called a unimodal function, or it could have one global maximum and multiple local maxima called a multimodal function. The example plotted in Figure 1.2 is unimodal whereas the plot shown in Figure 1.3 is multimodal.

1.3 Types of Optimization Problems

Optimization problems usually start with the definition of the problem such as minimization of cost, energy, resources used, or maximization of profit and quality. From this problem statement the objective function is formulated, either as a maximization or minimization function. The next step is to identify the constraints associated with the problem, which could be either equality or inequality constraints. The parameters associated with the objective function and the constraints have to be identified and their boundaries clearly stated. The optimum solution for the problem has to be found by searching, for which the search or solution space needs to be defined. The tentative location of the solution in the search space or the local region where the solution could possibly be found also has to be initially known, because that is the point at which the search for the optimum has to begin. If the information about the local region where the solution is likely to be found is not available, then the search for the optimum has to start from a random location in the search space. If the space is quite large and multidimensional, it is neither practical nor feasible to do an exhaustive search of the solution space. If there are no objectives in the problem but only constraints, then it is a *feasibility* problem. If the objective function and constraints are *separable* in the design variables, it is a *separable* optimization problem. Given any mathematical function of design variables and constraints, if it is possible to separate them in terms of the variables then it is a *separable* problem. Let a function of three design variables be defined as $f(X) = x_1 x_2 x_3$. Taking logarithm on both sides, log $[f(X)] =$ $\log(x_1 x_2 x_3) = \log x_1 + \log x_2 + \log x_3$. Since this is separable in terms of the three variables x_1, x_2, and x_3, it is a *separable* problem.

The optimization problems and techniques are categorized into several classes based on the characteristics of the objective function, the associated variables, and constraints.

Continuous and Discrete Optimization: When the objective function is a continuous function of design variables it is continuous optimization. This continuous function could be differentiable or non-differentiable. If the function is differentiable, the traditional or classical methods of optimization can be applied. $f(x) = 12x^3 + 5x^2 - 2x + 4$ is an example of a continuous function that is differentiable, and the traditional methods of optimization can be applied to find the minima or maxima of the function $f(x)$. If the function is non-differentiable, then other methods have to be explored, or they have to be transformed such that the classical methods can be applied. Such non-differentiable functions are usually not smooth and have sharp discontinuities where the derivatives do not exist. When the objective function is dependent on a set of independent discrete variables it is discrete or combinatorial optimization. Sometimes, the function could be a mix of continuous and discrete variables and it becomes a mixed optimization problem. An example of a combinatorial optimization problem is finding the minimum spanning tree of a connected

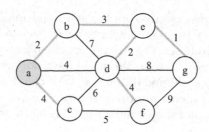

FIGURE 1.5
Minimum spanning tree.

undirected graph. The problem is to connect all the nodes of the graph such that the total weight of the edges connecting these nodes is minimum. In addition, there should not be any cycles in the minimum spanning tree. Kruskal's and Prim's algorithms are two greedy algorithms that are normally applied in finding the minimum spanning tree of a graph.

Figure 1.5 shows an example of a connected graph, and its minimum spanning tree is shown highlighted. Initially starting from node *a* of the graph, all the edges connecting this node are examined and the edge with the least weight of 2 is chosen and the node *b* connecting this edge is included in the tree. The edges connected to nodes *a* and *b* that are not yet included in the tree are examined, and the edge with a weight of 3 connecting node *b* to node *e* is included in the tree. This process is repeated until all the nodes in the graph have been covered. The algorithm for obtaining the minimum spanning tree is a greedy algorithm since it chooses the edge with the smallest weight at each step. The final optimized output is the set of edges whose total weight is minimum with all the nodes of the graph being connected and included in the tree. This is an optimization problem with the objective function being the total weight of all the edges in the minimum spanning tree and the constraints being that each node be included only once and no cycles are formed.

Deterministic and Stochastic Optimization: Deterministic optimization algorithms are consistent, and each time the algorithm is run, it produces the same result for the same input values. Stochastic optimization algorithms have some randomness associated with them and might produce different results for each run of the algorithm. An example for the deterministic optimization algorithm is hill climbing and other traditional gradient-based methods such as the simplex method. Genetic algorithm and particle swarm optimization are two famous examples of population-based stochastic optimization techniques. Stochastic optimization algorithms are heuristic or metaheuristic and have some inherent randomness that leads to the optimum or near-optimum solution in finite time.

Constrained and Unconstrained Optimization: When there are no constraints on the problem to be solved it is unconstrained optimization. When there are constraints related to the problem, it is constrained optimization. Let $f(x_1, x_2) = x_1^2 - 10x_1 + 5x_2^2 - 2x_2 + 3$ be the function to be optimized. The problem is to find the values of x_1 and x_2 that will maximize $f(x_1, x_2)$. If there is no restriction on the values that x_1 and x_2 can assume, it is unconstrained optimization. If it is stated that x_1 can only be a positive integer then it is constrained optimization. The constraints could be equality or inequality constraints. The constraint $x_1 \geq 0$ is an inequality constraint whereas $2x_1 + x_2 = 6$ is an equality constraint. Constrained optimization problems are more difficult to solve than unconstrained optimization problems.

Linear and Non-linear Optimization: If the objective function to be optimized is linear then it is linear optimization. If the objective function is non-linear then it is non-linear optimization. An example for a linear objective function is $f(X) = 2x_1 + 4x_2 - 7$, and a non-linear objective function is $f(X) = x_1^2 + 3x_1 - 2x_2^2 + 5$. Non-linear optimization problems are more difficult to solve than linear optimization problems. Similar to objective functions, the constraints also could be linear or non-linear, for example, $x_1 + x_2 = 6$ is a linear constraint and $x_1^2 + x_2^2 > 100$ is a non-linear constraint.

Single- and Multi-Objective Optimization: When there is only one objective function $f(X)$ to be optimized it is single-objective optimization. When there are multiple objective functions $\{f(X_1), f(X_2), ..., f(X_K)\}$ to be optimized it is multi-objective optimization. In multi-objective optimization there are multiple (K) objective functions and each one is a function of d number of design variables. In multi-objective optimization, there could be multiple functions with conflicting requirements. There will be no single design vector that will maximize (or minimize) all the objective functions. It might not be possible to satisfy all the constraints to get an optimum value for all the objective functions. It might be necessary to have tradeoffs when there are conflicting requirements between multiple objectives. A typical example with multiple objectives is manufacturing a car. One objective will be minimization of cost, the second will be minimization of fuel consumption, the third will be maximization of efficiency, the fourth will be minimization of resources used in manufacturing, the fifth will be minimization of emission of pollutants, and so on. Algorithms that work for single-objective optimization might not work directly for multi-objective optimization. One method to solve such multi-objective optimization problems is to construct a single-objective function using a weighted combination of the multiple objective functions such as:

$$f(X) = c_1 f(X_1) + c_2 f(X_2) + ... + c_K f(X_K) \tag{1.4}$$

where $c_1, c_2, ..., c_K$ are the weighting coefficients.

A better method to take care of these conflicting requirements is that the optimal values of all the objective functions lie on the Pareto Optimal Front. The Pareto Optimal Front theory states that it is not possible to get the maximum value for all the multiple conflicting objectives of a problem simultaneously. What is stated here for a maximization problem applies equally well for a minimization problem. The optimal method to solve this multi-objective problem with conflicting requirements is to ensure that the solutions to the multiple objective functions lie on the Pareto Optimal Front. When the fitness value of one objective function is increased (improved), the fitness value of another objective function will be decreased (degraded); that means it is not possible to get the maximum value for all the objective functions simultaneously with one solution vector. If a solution vector X_i lies on the Pareto Optimal Front, it yields a vector of fitness values for all the objective functions of the problem. Comparing these fitness values, it occurs that if this vector X_i yields maximum fitness for objective function $f(X_i)$ it might not be the case for the function $f(X_j)$. Another solution vector X_r which is also on the Pareto Optimal Front will produce the maximum value for the objective function $f(X_r)$ but not for $f(X_i)$ or $f(X_j)$. Therefore, any solution vector X for a multi-objective optimization problem should lie on the Pareto Optimal Front for the solutions to be non-dominated or Pareto Optimal.

1.4 Examples of Optimization

These algorithms have one thing in common, that is, they all strive to produce the best solution for a given problem within a limited set of resources. The problem can be a complex engineering design such as a CNC Machine, a process plant like paper manufacturing, pattern recognition, image classification, or an NP-hard problem (non-deterministic polynomial time) in computer science like the traveling salesman problem (TSP), and the limitations on the resources are the constraints within which the best solution for the problem is to be attained. It can be expressed in terms of mathematical equations that could be solved using different methods suitable for the problem. One of the most important evaluation criteria for these algorithms is the time, space, and computational complexity incurred in arriving at the optimum solution.

Paper Manufacturing

In a process plant like paper manufacturing, many parameters will be involved in producing the paper. The objective function could be a mathematical function that has to be maximized and indicates the quality of the paper produced. There are several processes involved in paper manufacturing such as pulp extraction, the right blend (proportion) of chemical additives, fixing temperature and/or pressure, drying, bleaching, etc. Choosing the appropriate values for the parameters involved in the various processes plays a crucial role in the color, thickness, and quality of the paper produced. This is an optimization problem where the parameters of the paper manufacturing process are equivalent to the decision variables $X = \{x_1, x_2, ..., x_d\}$ in the search space and the function $f(X)$ is the fitness or quality function. The value of the fitness function has to be proportional to the paper quality produced. The global optimum of the objective function will be attained when the function has the highest value and the paper produced is of the best quality.

Pattern Recognition

Consider the pattern recognition problem where the objective is to identify and classify objects in an image with 100% accuracy. Every object such as jasmine, rose, car, lorry, bike, building, human face, and fingerprint present in an image has its own set of features that enables recognition and classification. Flowers have features like the number of petals, length and width of the petals, color, diameter of the flower, etc. Similarly a fingerprint has features like ridges and minutiae. The ridge and minutiae patterns on the fingerprint enables classification and recognition.

Feature Reduction

Each object or pattern has its own unique set of features that enables identification. When the number of features of an object is higher it becomes impractical to extract all the features, compare it with the existing template, and then identify the object. So to circumvent the problem, a technique known as feature reduction is applied. This is the technique of selecting a minimum number of features from the entire set, for classification of objects with a reasonable percentage of accuracy. Minimizing the number of features is nothing but the selection of a subset of features, thus reducing the computational complexity in searching for the optimum solution. It is equivalent to minimizing the value of dimension d in the hyperspace. Once d is fixed, the next step is to find the value of the feature set $X = \{x_1, x_2, x_3, ..., x_d\}$ so that $f(X)$ attains the global maximum or minimum value, as the case may be. If $f(X)$ represents classification accuracy then it will be maximized whereas if $f(X)$ is a classification error then it has to be minimized.

1.5 Formulation of Optimization Problem

The first step in any problem solving is definition of the problem statement. Once the statement is clear, the objectives and constraints have to be outlined. The objective(s) of the problem have to be mathematically formulated as an objective function that is to be either maximized or minimized. The next step is to identify the limitations or boundaries of the design variables as equality or inequality constraints and write mathematical equations for them. When the equations are linear and differentiable, with few parameters, the standard classical methods can be applied. When the equations are non-linear and complex or if the objective function and/or constraints have discontinuities and are not differentiable, the classical methods could become intractable, necessitating the use of evolutionary or metaheuristic algorithms. The nature-inspired algorithms that are basically metaheuristic, population-based search algorithms will be a better choice for such complex problems. Many of the heuristic methods use a greedy criterion in accepting a new solution component that might evolve from the previous solution(s), if it is either going to increase the fitness value or decrease the cost. There is also a possibility of converging at a local optimum for such greedy algorithms, which could be overcome by the metaheuristic algorithms that search in parallel and have good diversity properties.

Although several mathematical techniques exist for solving optimization problems, the computational complexity involved might be too high in some of the cases. Certain approximations might reduce the search and computational complexity and help in arriving at the near-optimum solution in finite time for practical applications. When the number of design variables upon which the objective function is dependent is large, the dimensionality of the problem increases. This can be reduced by dimensionality-reduction methods based on the specific optimization problem. Sensitivity analysis of the objective function with respect to the parameters, i.e. change in the objective function value with changes in the design parameters could be done to improve the parameter settings and hence the performance of the algorithm. As another alternative, multilevel optimization could be applied for large complex problems. When the number of design variables and/or constraints becomes large, the optimization problem becomes unmanageable or impractical. Multilevel optimization involves breaking down an optimization problem of large size or dimensions into smaller sub-problems that can be optimized easily. The smaller sub-problems are linked to put together a solution for the larger problem. Yet another approach is parallel processing, where the sub-problems can be run independently in multiple parallel computers to speed up the optimization process.

The problem is to be defined clearly and the physical principles governing the system to be optimized are to be understood thoroughly. The limitations under which the solution is to be found are formulated as constraints. For example, the number of persons cannot be a fractional number, the resistance value cannot be negative, and so on. These restrictions on certain parameters can be formulated as equality or inequality constraints. Based on the mathematical equations, the problem can be categorized as linear, non-linear, integer, quadratic, etc. When probabilistic variables are involved, it is stochastic programming. Necessary and sufficient conditions for optimality have to be identified and the optimum solution must satisfy these conditions. The best solution among all those available is to be selected based on some criterion. The optimum solution gives a maximum performance measure as compared to other non-optimal solutions. Optimization methods can also be classified as direct and indirect. Direct methods make use of objective functions and constraint equations whereas indirect methods use properties of functions. Most of these approaches are suitable for programming on a computer, and hence scaling or transformation of variables can be accomplished easily.

1.6 Classification of Optimization Algorithms

The optimization algorithms are classified into the traditional or classical methods, evolutionary algorithms, and swarm intelligence algorithms.

Classical Methods: The classical optimization algorithms are applicable to the traditional continuous optimization functions that are differentiable. The traditional methods involve computation of the first- and second-order derivatives of the objective function in order to find maxima or minima. The solution obtained could be local or global optima but the ultimate goal of optimization is to find the global optimum. They are mostly gradient-based methods and provide deterministic solutions to optimization functions within a continuous search space. The simplex method, linear and non-linear programming, Newton's steepest descent method, Lagrangian method, integer and dynamic programming, and the Kuhn–Tucker conditions are famous classical methods, to name a few. Steepest descent is one of the classical optimization techniques for finding the minimum of a unimodal differentiable function. Figure 1.6 illustrates the trajectory of the steepest descent method. In the classical descent methods, the step size plays a major role in approaching the minimum of a unimodal function. The brute force method is a single point direct search method that searches for the optimum point starting from a single point in the search region that is bounded. The search proceeds in steps, and the step size plays a major role in convergence of the algorithm. When the number of parameters is large, the brute force method suffers from the *curse of dimensionality*.

The random walk method overcomes or circumvents the *curse of dimensionality* by searching from randomly generated points in the search space of the objective function.

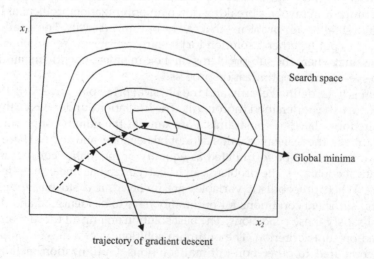

FIGURE 1.6
Illustration of gradient descent.

The Hooke–Jeeves method is a direction or pattern search since it starts the search from a base point and has separate step size for each coordinate direction. The trial search points are compared, and if a move is found to be better, then further moves are made in the same direction. Thus the step sizes are adaptive in this method which makes it perform better than the earlier two methods. When the functions are multimodal, the starting point of the search plays a vital role in finding the global optimum since there is more than one optimum (local as well as global) point. There are other methods that are called multi-start since they search the solution space starting from multiple different initial points for each search. The search could be direct or derivative-based. Clustering could be applied where clusters of sample points are formed and one point within the cluster serves as the initial point for the search. This is limited to problems with smaller numbers of parameters since it is computationally intensive. An overview of the classical algorithms has been discussed in the following chapter.

Evolutionary Algorithms: Evolutionary algorithms are a class of optimization algorithms that are based on the biological processes of evolution. As the name implies, evolutionary computational methods are designed on the principles of evolution. They are based on the Darwinian theory of *survival of the fittest* and use selection, mating, reproduction, crossover, and mutation operators similar to the biological processes of evolution. Genetic algorithm (GA), genetic programming (GP), and differential evolution (DE) are three of the classical popular evolutionary algorithms that have been discussed in this book. *Darwin's finches* belong to a group of passerine birds that have around 15 different species within their group. They are diversified in the shape, size, and function of their beaks. Long-term study has shown evolutionary changes in their beaks. They are popularly known as *Darwin's finches* because Charles Darwin found and collected them from the Galapagos Islands and they formed the basis for Darwin's theory of evolution and natural selection. Figure 1.7 shows the *Darwin's finches* that played an important role in Darwin conceiving his theory of evolution and are a typical example of evolution.

Additionally, other evolutionary algorithms available are gene expression programming (GEP), evolutionary programming (EP), and evolutionary strategies (ES). Charles Darwin developed the theory of evolution on which these algorithms are all built upon, and they are all population-based. The fundamental goal of the theory or algorithm is survival of the species.

The transfer of a genetic program (genotype) or code from an individual to its progeny is reproduction. When there are mutations, there are differences in the transferred genetic code that are either advantageous or disadvantageous. When there are more individuals than resources, there is competition and the fitter ones tend to survive and reproduce whereas the weaker ones perish. The genotype carries genetic information from the parent to offspring with all the experience undergone by the parent so far. Phenotype is the set of characteristics or properties of the population, which is a manifestation (behavioral expression) of the genotype in a specific environment. In natural evolution, there is non-linear mapping between genotype and phenotype and it is complex. The algorithms that are developed under the evolutionary framework are broadly referred to as evolutionary algorithms. The evolution of the population is determined by the parents, offspring, and the operators, and the algorithms differ from each other based on this combination. Let P be the population size of parents, C be the number of offspring or children, and after recombination the population size is $P + C$. This has to be resized or scaled down to the original size, and the operators determine the mode of resizing. The fitness of every member of the population is evaluated by the objective function, and the overall fitness is determined by the individual fitness values.

1. Geospiza magnirostris.   2. Geospiza fortis.
3. Geospiza parvula.   4. Certhidea olivasea.

FIGURE 1.7
Darwin's finches. [Author: John Gould (Public Domain).]

In evolutionary strategies the search is in parallel (multi-start) from multiple points in the search space. The initial population is created randomly, and parents are selected for recombination to create a population of children. The children are mutated to change their properties or characteristics; either all of them undergo mutation or it is done selectively. The next-generation population members are chosen from the total population consisting of parents and children. This resizing of the population is done based on the fitness values of the individual members.

Swarm Intelligence Algorithms: Computational intelligence (CI) is a subset of machine learning and artificial intelligence that collectively refers to algorithms with intelligence built into them. Most of the evolutionary algorithms and swarm intelligence algorithms have computational intelligence built into them. Some of the popular computational intelligence algorithms are neural networks, fuzzy logic, genetic algorithm, differential evolution, particle swarm optimization, firefly algorithm, ant colony optimization, and other swarm-based algorithms. Nature-inspired (NI) algorithms are developed based on the study of natural processes that could be physical, chemical, or biological and the behavior of animals, birds, and insects. They are mostly population-based and utilize their collective swarm intelligence in arriving at the optimum solution. Swarm intelligence (SI) algorithms have been developed based on the study of swarm behavior, and these swarms possess computational intelligence. Swarm behavior refers to the collective behavior of a group of insects, birds, fish, and animals, and swarm intelligence is the intelligence exhibited by the entire group as a whole. The individual members of the swarm follow simple rules that lead to collective and productive outcomes for the benefit of the entire swarm. They are adaptive and flexible, have good perception capabilities, and live in

FIGURE 1.8
Weaver ants making an emergency bridge between two plants. (Author: Rose Thumboor, CC BY-SA 4.0. https://creativecommons.org/licenses/by-sa/4.0/deed.en.)

the environment utilizing the available resources efficiently. They interact within their group as well as with the environment and behave in a self-organized manner. Most of the swarms have a hierarchy within their group and apply principles of division of labor. Their collective intelligence helps them in foraging for food, defending against predators, and societal interactions such as communication among the members of their group for the benefit of the entire swarm. Such studies have motivated the research towards development of swarm intelligence algorithms that are able to solve complex problems efficiently in finite time. Figure 1.8 shows a group of weaver ants making an emergency bridge between two plants that is a manifestation of group dynamism and collective intelligence.

Many of the classical optimization algorithms require the objective function to be a continuous function whose derivative exists. If the function is discrete in nature or if there are numerous parameters in the problem, these algorithms fail. They are not suitable for dealing with functions whose derivatives do not exist or for discrete optimization problems such as in image processing. In such cases, computational intelligence algorithms are best suited to provide the solution. They are mostly designed for searching a solution space trying to find the best solution or a close approximation. This involves some randomness in the algorithm which in itself increases the efficiency of such methods. Searching a solution space is best done using a swarm or a group of particles, and these serve as the basis for nature-inspired algorithms. They are all population-based and reduce the time of searching to find the optimum solution. SI-based techniques are mostly iterative, and they remember the past history in order to find a better solution than the previous best. Remembering the previous solutions involves memory, and many of the SI algorithms do have memory, either short-term or long-term. The algorithm stops either when a stopping criterion is attained or when the predefined maximum number of iterations is reached. The two most important components in swarm intelligence algorithms are intensification and diversification. In intensification, the search is intensified in a particular area near the previous best solution, whereas in diversification, the search is conducted over a wider area that was previously unexplored, looking for better solutions.

1.7 Traveling Salesman Problem and Knapsack Problem

Optimizing a cost function that is defined on a set of independent variables is a combinatorial optimization problem since it involves finding the right set of independent variables that maximizes or minimizes the function. The problems which generally fall into this category can be divided into two classes – those which are easy to solve and take up less time (can be solved in polynomial time) and those which take up a large amount of time and are practically infeasible to solve, called NP-hard problems. The most famous problem in computer science under the category of NP-hard is the traveling salesman problem (TSP), and other similar problems in this category are the graph partitioning (coloring) problem, job scheduling problem, and network routing problem, to name a few. These combinatorial optimization problems could be solved by the classical optimization algorithms in an infinitely long time or by the heuristic/metaheuristic optimization algorithms in finite short time that produces an optimum solution that closely approximates the actual solution.

The TSP and the knapsack problem are two typical famous examples in computer science for discrete combinatorial optimization. In the TSP, the objective function is the total distance of the Hamiltonian tour and depends on the set of cities visited and the total distance covered which is the sum of the distance of the edges connecting the cities, and in the knapsack problem the combination of the weights of the items loaded into the knapsack and their total cost determines the objective function value. Consider the traveling salesman problem which is a typical combinatorial optimization problem in computer science that has been found to be intractable or NP-hard. There are a set of cities that are interconnected. The cities are represented by nodes in a graph and the connections between the nodes are edges. Each edge has a value associated with it that could represent distance or cost. The problem is for a traveling salesman to visit each of these cities once and only once (without retracing) and go back to the starting city traveling a total minimum distance and hence incurring the minimum cost. The path in the graph traced by the traveling salesman is called the Hamiltonian tour. This is illustrated in Figure 1.9 which shows one possible Hamiltonian tour in the graph.

Consider the *knapsack problem* in computer science which is another typical combinatorial optimization problem. There is a knapsack or rucksack and a set of items each with a value and a weight. The items are represented by x_k with a weight of w_k and value v_k, where $k = 1, 2, ..., N$, assuming there are N number of different items. The objective is to fill the knapsack with as many items as possible so that the total value of the items is maximum. The items can be included as a whole or in fractions, such as c_1x_1 where $c_1 = 1$ is one instance of the item x_1, $c_1 = 2$ is two instances of the item x_1, whereas $c_1 = 0.5$ is one-half of the item x_1. The constraint is that the maximum weight of all the items put in the knapsack should not exceed its capacity of W_{max} kg. In this problem, the objective function is the total value of all the items in the knapsack which can be formulated as the weighted sum of the values of each item put in the knapsack. The constraint is that the sum of the weights of all the items put in the knapsack should not exceed the limit of W_{max} kg. This is a multi-variable, constrained optimization problem with a single objective function that is

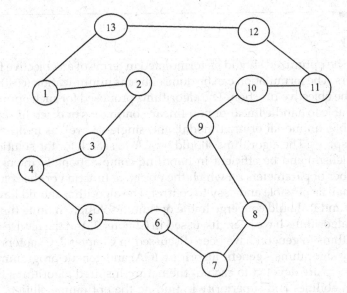

FIGURE 1.9
Hamiltonian tour for TSP.

to be maximized. This is illustrated in Figure 1.10. The objective function is a cost function that is to be maximized, given by:

$$f(X) = c_1 v_1 + c_2 v_2 + \ldots + c_N v_N \quad (1.5)$$

The constraint is:

$$c_1 w_1 + c_2 w_2 + \ldots + c_N w_N \leq W_{\max} \quad (1.6)$$

where the coefficients $c_1, c_2, \ldots, c_N$ are either integers or fractional numbers.

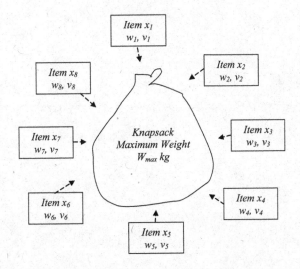

FIGURE 1.10
Knapsack problem.

1.8 Summary

Any problem to be optimized should be formulated in terms of an objective function and a set of parameters. The formulation can be done either as minimization (cost) or maximization (profit) of the objective function. The algorithms proposed for such optimization problems should be able to handle linear or non-linear, continuous or discrete, differentiable or non-differentiable, unimodal or multimodal, and single- as well as multi-objective functions or search spaces. The algorithms should be able to search for the solutions in parallel (inherent parallelism) and be efficient in handling computationally intensive cost functions. The number of parameters on which the objective function or the solution depends should be as small as possible and easy to control. The algorithm should have a faster rate of convergence, and it should converge to the optimum solution in finite time.

The classical algorithms have been discussed in Chapter 2, and the nature-inspired optimization algorithms in general have been discussed in Chapter 3. Chapters 4 and 5 cover the evolutionary algorithms – genetic algorithm (GA) and genetic programming (GP). The rest of the chapters are devoted to each of the nature-inspired algorithms that are popular, with proven abilities and superiority in finding the optimum solution. The standard benchmark functions used for testing and comparing the performance of these algorithms and typical applications for evaluating these algorithms have been outlined in Chapter 19. The book concludes with a summary in Chapter 20.

2

Classical Optimization Methods

2.1 Introduction

The goal of optimization is to get the best possible output from the available resources within a set of constraints. The output could be maximization of profit/efficiency or minimization of cost/resources utilized. Maximization and minimization are interchangeable since one can be converted to the other and *vice versa*. Optimization is applicable to diverse fields such as engineering design, network routing, job scheduling, communications, computer science, economics, business management, and a host of other complex applications. Several optimization algorithms have been developed with each algorithm suitable for a particular class of problems. The effectiveness of the algorithm is determined in terms of time, space, and computational complexity. The convergence rate of the algorithm and the computational resources utilized play a major role in the algorithm efficiency and optimality of the solution obtained for the problem. The requirement of quantitative results for the optimization problem necessitates writing mathematical equations for the objectives and constraints. This in turn requires mathematical techniques for solving the equations which might have some initial conditions and bounds or constraints on variables.

Operations research is the branch of optimization concerned with the mathematical analysis techniques required for producing the optimum output [1]. It is a discipline that deals with the analytical methods that aid in complex decision-making during problem-solving. Operations research involves construction of mathematical models wherein computer programs can be written to solve problems that have been modeled within a mathematical framework. When the problems involve stochasticity, statistical programming techniques need to be employed. The probability distributions of such stochastic variables have to be known in advance in order to fit the models correctly. The numerical programming techniques are the classical methods of optimization that have been developed since the early 1940s. The earliest developments were made by Cauchy, Newton, and Lagrange, and since then several techniques have been proposed. The advances were made possible by the parallel development of high-speed computing technology on which these algorithms were programmed. The methods discussed in this chapter are quite popular and have been used successfully in solving constrained as well as unconstrained optimization problems.

As an addition to these traditional methods, the evolutionary optimization algorithms were developed in the 1960s. The pioneering development in the class of evolutionary algorithms was made by the invention of the genetic algorithm, and since then this category of algorithms has grown by leaps and bounds. The application of computational intelligence in these algorithms was first done with the invention of neural networks. The inception of evolutionary algorithms paved the way for development of search algorithms using populations of agents that search for the optimum in parallel. This inherent parallelism in these algorithms has made them more efficient than their classical counterparts. One major difference is that the classical algorithms are deterministic whereas the evolutionary search algorithms are non-deterministic. The randomness incorporated into the search improves the diversity of the algorithm and enables the algorithm to jump out of local optima. This increases the efficiency of the algorithms in the search, and their rate of convergence on the global optimum is higher.

2.2 Mathematical Model of Optimization

The solution to the optimization problem commences with the definition of the problem statement with clearly defined objectives. The constraints and parameters associated with the problem need to be stated lucidly. The mathematical equations for the objective(s) and constraints of the problem have to be formed. The initial values for the functions, if any, and the bounds on the parameters have to be identified. The objective function value is an indication of the quality of the solution attained which can be framed for either maximization or minimization. A maximization objective function can be converted to a minimization function by taking the negative of the function and *vice versa*. Similarly the constraints could be written as equations of either the equality or inequality type. Depending on the number of variables upon which the objective function depends, the function becomes multidimensional; typically the dimension is d.

Let the objective function be given by,

$$f(X) = f(x_1 \ x_2 \ ... \ x_d) \tag{2.1}$$

Let the constraints of the problem be given by,

$$g_i(X) = 0, \quad i = 1, 2, \ ..... \ P$$

$$h_j(X) \geq 0, \quad j = 1, 2, \ .... \ Q \tag{2.2}$$

$$or \ h_j(X) \leq 0, \quad j = 1, 2, \ .... \ Q$$

In the case of multi-objective optimization problems, the number of objective functions is assumed as K and the multiple objective functions are given by,

$$f(X_k) = f(x_{k1}, \ x_{k2}, \ ......, \ x_{kd}), \qquad k = 1, 2, \ .....K \tag{2.3}$$

These functions could be of different types such as continuous or discrete, differentiable or non-differentiable, deterministic or non-deterministic with some random parameters inculcated into them. The traditional algorithms are usually applicable if the function is continuous and differentiable. When the function is constrained, it is a constrained programming problem whereas if it is unconstrained, it is an unconstrained programming problem. If there is only a single objective to be optimized, it is single-objective optimization, whereas if there are multiple objectives, it is multi-objective optimization. The function could be concave or convex as determined by the contours of the surface plot of the function. Concave functions have one maximum whereas convex functions have one minimum. A problem is said to be feasible if there exists at least one set of variables that satisfy the objective function and the constraints. A problem is infeasible if it is not possible to find at least one set of design variables that satisfy the objective(s) and the constraints. If the problem is infeasible the solution set will be empty. For an unbounded problem there is no optimal solution since it is always possible to find a solution that is better than the existing solution to the problem.

Based on the nature and characteristics of the objective function, constraints, design variables, and any other parameters associated with the problem, the techniques for solving the classical optimization problems can be categorized as follows:

- Linear programming
  - Simplex method
  - Revised simplex method
  - Kamarkar's method
  - Decomposition principle
  - Duality theorem
  - Transportation problem
- Non-linear programming
  - Quadratic programming
  - Geometric programming
  - Kuhn–Tucker conditions
- Dynamic programming
- Integer programming
- Stochastic programming
- Lagrange multiplier method

These traditional methods have been discussed in the following sections with simple examples where applicable.

2.3 Linear Programming

Linear programming is a numerical programming technique that is designed to solve optimization problems where the objective function and constraints (equality and inequality) are expressed as linear equations (functions) of the decision variables. An example of the mathematical modeling of a linear programming problem is:

$$\text{Minimize } f(X) = 2x_1 + 4x_2 - 5x_3 + x_4$$

subject to the constraints:

$g_1(X): x_1 + x_2 = 6, \ g_2(X): 2x_1 - x_4 = 8, \ g_3(X): x_3 + x_2 < 25.$

In general, there are d number of variables and m number of equations (constraints) in these variables with $d \geq m$. The inequality constraints can be converted to equality constraints by the inclusion of additional terms in the constraint equations.

Usually in linear programming problems, the decision variables are positive numbers like $x_1 \geq 0$, $x_2 \geq 0$, $x_3 > 0$, $x_4 > 0$. The set of linear equations (objective function and constraints) may also be rewritten in matrix notation. The feasible solution space will be convex with one global optimum (minimum). When the number of decision variables is more than two, the solution space becomes a hyperplane in d dimensions. George B. Dantzig formulated the linear programming problem in 1947 and invented the *simplex method* to solve linear programming problems [2]. This was a major milestone development that led to several applications such as optimal allocation of resources, scheduling, maximizing production in petroleum refineries, manufacturing plants, and optimal design of structures. It had far-reaching implications in the fields of mathematics, economics, computer science, industry, military, and various other diverse fields. This invention by George B. Dantzig had tremendous impact on solving complex problems efficiently, and he was awarded the US President's National Medal of Science in 1976.

2.3.1 Simplex Method

The *simplex method* was one of the earliest methods developed for solving linear programming problems and it is efficient and has become very popular. In the simplex method, the search for the optimum solution in the feasible space initially starts from a vertex and proceeds along the edges of the space to find better solutions. A basic set of feasible solutions is generated, and one solution from this set is chosen. If this solution is suboptimal, its neighboring solution is chosen and checked to see whether it is better than the previous one. If the second one is either the optimum or better than the previous one, this solution replaces the previous one. This process is repeated among all the available solutions in the set until the optimum solution is found.

2.3.2 Revised Simplex Method

Revised simplex method is a variant of the original simplex method that has reduced storage requirement and computational time. The simplex method requires large memory capacity in the computer to store the several variables and the constraints involved when the problem has a higher number of dimensions. It also takes up more time for computations. In the original simplex method, a new table has to be computed and stored in each iteration of the algorithm. This takes up more memory and time and most of this stored information is not used in every iteration. This memory requirement is reduced in the revised simplex method by storing the basis of the matrix representing the constraints. The required quantities are computed from the inverse of the current basis matrix. This makes the computations more efficient.

2.3.3 Kamarkar's Method

Kamarkar's method was developed in 1984 for solving large-scale linear programming problems [3]. The algorithm proposed solves the linear programming problems efficiently with reduced time complexity. The simplex method searches for the optimum solution along the boundaries (vertices) of the feasible solution space whereas Kamarkar's method searches for the optimum solution in the interior of the feasible solution space. In bigger problems with a large number of vertices, it is computationally intensive to search along the vertices for the feasible solution. Kamarkar's method transforms the solution space so that the optimum lies near the center of the bounded space. The mathematical equations for the linear programming problem are transformed, and the global minimum for the problem is required to be necessarily zero so that this technique can be applied. Kamarkar's method is approximately 50 times faster than the simplex method in solving large-scale problems.

2.3.4 Duality Theorem

With every linear programming problem there is associated another problem. The first is called *primal* and the second is called *dual* and their names can be interchanged. They have related properties such that from the solution of one problem, the solution to the other can be obtained. If the *primal* problem can be expressed with one set of linear equations that can be represented in matrix form, the *dual* problem can be expressed by transposing the rows and columns of the matrix of the *primal*. Moreover, the inequalities have to be reversed and the maximization has to be replaced with minimization.

The *duality theorem* states the following:

- Dual of the *dual* is *primal*.
- If the optimum solution to one of them is known, then the optimum solution of the second one will be better than or equal to the optimum solution of the first one.
- If *primal* and *dual* have optimum solutions, the global maximum value of the *primal* function is the global minimum value of the *dual* function.
- If one of them has an unbounded solution, then the optimum solution to the other one is also not feasible.

The set of problems in linear programming called *primal* possess properties that define their *dual*. The *primal* and *dual* are closely related with a symmetric relationship, such that if the solution to the *primal* is known, the solution to the *dual* can be easily obtained and *vice versa*.

Let the *primal* problem be expressed as a set of linear equations with an objective function to be maximized and constraints with *greater than or equal to* inequalities. Then the equations for the *dual* of the problem can be expressed by replacing the *primal* objective function with an equivalent minimization function and constraints with *lesser than or equal to* inequalities. The matrices of the *primal* and *dual* are transpose of each other. The *dual simplex method* and the *primal–dual method* are widely used for solving such problems. *The dual simplex method* [4] was developed by Lemke (1954), and the *primal–dual method* [5] was developed by Dantzig, Ford, and Fulkerson (1956).

Consider the example given below:

Primal problem:

Maximize $f(X) = \sum_{i=1}^{d} a_i x_i$ subject to the following constraints:

$$\sum_{i=1}^{d} c_{i1} x_i \geq b_1 \quad \sum_{i=1}^{d} c_{i2} x_i \geq b_2 \quad \ldots \quad \sum_{i=1}^{d} c_{im} x_i \geq b_m$$ and the set of variables $\{x_i\}_{i=1}^{d}$ constrained to be positive.

Dual of the above problem is:

Minimize $g(Y) = \sum_{j=1}^{m} b_j y_j$ subject to the following constraints:

$$\sum_{j=1}^{m} c_{j1}y_j \le a_1 \quad \sum_{j=1}^{m} c_{j2}y_j \le a_2 \dots \sum_{j=1}^{m} c_{jd}y_j \le a_d$$ and the set of variables $\{y_j\}_{j=1}^{m}$ constrained to be positive.

The optimal values of the *primal* and *dual* problems are equal. The computation time of the problem increases with the number of constraints rather than with the number of variables. If the *dual* of the problem has a lower number of constraints then it can be solved more efficiently than the *primal* yielding the same solution.

2.3.5 Decomposition Principle

The *decomposition principle* was proposed by George B. Dantzig and Philip Wolfe in 1960 for solving large-scale problems which are difficult to solve in linear programming [6]. When the number of variables involved in a problem is high and there are several constraints, the problem becomes complicated and unwieldy. The *decomposition principle* can be applied when the larger problem has a special structure that makes it possible to decompose into smaller sub-problems which can be solved independently. Stated briefly, the principle involves dividing a large problem into multiple smaller problems that are easier to solve. The solutions for the sub-problems are put together to obtain the solution for the bigger problem. This decomposition technique has made computations for solving large problems less complicated and more efficient.

2.3.6 Transportation Problem

The transportation problem was formulated by F. L. Hitchcock [7, 8] in 1941 as a special class of linear programming problems. The basic problem is stated as follows. There are W number of warehouses and R number of retail outlets. The problem is to transport a certain quantity of an item from each of the warehouses to the retail outlets with the constraint of a maximum supply from each warehouse and maximum demand at each retail outlet. The total cost of transportation from each warehouse to each of the retail outlets will be the objective that is to be minimized. This type of problem is found to have a special mathematical structure that is convenient and easy to solve. Utilizing the properties of the matrices associated with the transportation problem, the transportation technique for solving the problem has been devised. It is computationally less expensive compared to other linear programming problems. This transportation problem structure appears in other applications such as job scheduling, finding the shortest path in a network, and so on.

2.4 Non-Linear Programming

This is an optimization technique where the objective function and either all or some of the constraints are non-linear equations of the design variables. Several algorithms have been proposed for the solution of such non-linear programming problems, a few of which are discussed below. Some of the techniques are applicable to both linear as well as non-linear optimization problems.

2.4.1 Quadratic Programming

When the objective function of the problem is a quadratic function of the variables with linear constraints it is quadratic programming. This is a special case of non-linear programming. In addition, if the optimization problem is minimization of the objective function, it is also convex. Therefore there will be only one minimum in the feasible solution space which is the global minimum. Since the objective function has quadratic terms it becomes a non-linear programming problem which is an extension of the linear programming technique. The simplex method for solving quadratic programming problems was proposed by Philip Wolfe in 1959 [9] and can be applied for solving the class of optimization problems that have a quadratic function with linear inequality constraints. The constraints are assumed as inequalities and linear function of variables. This method can be applied to certain types of problems like finding the least squares fit to a given set of data (regression) and finding the quadratic approximation to the minimization problem of a convex function (convex programming).

2.4.2 Geometric Programming

Geometric programming is an optimization method for solving non-linear programming problems. Duffin, Peterson, and Zener developed the geometric programming technique in 1967 [10]. The engineering design problem has to be stated in terms of polynomials. The objective function as well as the constraints have to be in polynomial form [11] and the algorithm finds the minimum of the function. The optimal minimum value can be obtained as the solution without the necessity of finding the values for the design variables. This is one of the advantages of geometric programming. Another advantage is the ability to reduce the problems into a set of simultaneous linear algebraic equations [12]. The disadvantage is having to express the problem and its constraints in terms of polynomials. Consider the example given below where the design problem is formulated as a polynomial:

Minimize $f(X) = 2x_1 + 3x_2 - x_1^2 + 2x_1x_2 - 5x_2^2$ where the design variables are x_1 and x_2. If it is an unconstrained optimization problem there are no associated constraints. If it is a constrained optimization problem, the constraints (either equality or inequality) may be additionally expressed as:

$$g_1(X): x_1 + x_2 < 10 \text{ (inequality constraint)}$$

$$g_2(X): x_1 - 4x_2 = 0 \text{ (equality constraint)}$$

In general, the objective function and the constraints are d-dimensional and there could be multiple constraints. For such problems the general equations are expressed as:

Minimize $f(X), X = \{x_1, x_2 \ldots \ldots x_d\}$ subject to the constraints:

$$g_j(x) > 0, j = 1, 2, \ldots . J \text{ (inequality constraint)}$$

$$h_k(x) = 0, k = 1, 2, \ldots \ldots K \text{ (equality constraint)}$$

The geometric programming problem can be solved using either *differential calculus* or *arithmetic-geometric inequality*. The solutions to geometric programming problems can be obtained efficiently provided the design or analysis is stated in terms of mathematical equations appropriately. This can be extended to problems from small scale to large scale. Geometric programming also belongs to the class of *convex optimization problems*.

Kuhn–Tucker conditions are the conditions to be satisfied at a minimum point for a constrained optimization problem [13]. They include equality as well as inequality constraints. They are necessary conditions, but they might not be sufficient for the point to be a minimum in non-linear programming. For convex programming problems, *Kuhn–Tucker* conditions are necessary and sufficient. Kuhn and Tucker first published the conditions in 1951.

2.5 Dynamic Programming

Dynamic programming is a method to solve optimization problems that require decisions to be taken as a sequential flow through the problem. The decisions could be required at different levels of the system. This technique of solving such problems was developed and presented by Richard Bellman [14] in 1954. Physical systems can be modeled as finite state machines where the set of parameters of the system determines the system state. The system is modeled or designed as going from one state to another state in time sequence. The system changes state based on a decision taken at that instant of time. Finally, a function which depends on these parameters attains a value that is determined by the sequence of decisions taken and hence change of states (transformation of state variables) of the system. This value attained by the function could be maximum or minimum but it should be optimal. Examples of such systems could be the production line in a manufacturing plant or maintenance of equipment and consumables in a factory.

Considering the set of all possible decisions that could be taken at different instances of time (in time sequence) the change of state for each of these decisions affects the consecutive states. This in turn affects the final value of the objective function or some output equivalent to the function value. When all of these possibilities are considered the search space for the optimal solution becomes huge. To reduce the problem dimension so that the solution can be obtained reasonably in finite time, it is reduced to a sequence of decision problems without taking into account the effect of every decision at a later instant of time. Therefore, the K-stage problem is broken down into K single-stage problems that can be solved much more efficiently than the original K-stage problem. The overall solution should be same as that of combining the K different solutions.

The initial conditions and the decisions taken at the earlier stages of the system will not impact the decisions taken at later stages. This principle breaks the problem down into K different sub-problems which are simpler to solve. Finally, these K solutions can be combined to obtain the overall optimum solution. The system undergoes a transformation with every decision taken (at every stage), and the subsequent decisions should result in the optimum output. When there is a multistage decision problem involving K variables, it can be broken down into K single-stage problems, each being modeled as a single-variable problem. The solution to the K single-variable problems can be combined to obtain the solution for the K-variable problem. Any of the methods available can be used to solve these single-variable problems. Dynamic programming can be applied for problems wherein the variables and functions are continuous, discontinuous, discrete, differentiable, non-differentiable, etc. It can also account for stochastic variations and is found to be suitable for complex design and analysis of engineering problems.

2.6 Integer Programming

When solving engineering design problems the variables or the function might attain real values that could be positive or negative with integer and fractional parts. If the variable represents length or weight, then fractional values such as 4.89 m or 78.98 kg have meaning. If the variable represents people or objects, then fractional values such as 5.7 or 0.77 are meaningless. The values could be rounded off, but this might affect other values and hence the constraints of the problem might not be satisfied. The value of the objective function could also be different from that of the optimum because of rounding off to nearest integer values of the design variables.

Example

Minimize the objective function $f(X) = 3x_1 + x_2^2 - 6x_3$ subject to the constraints
$$x_1 \geq 0, \quad -5 \leq x_2 \leq 5, \qquad x_1, x_2, x_3 \text{ are necessarily integers.}$$
When all the variables in a problem are restricted to take only integer values, the optimization problem becomes an *integer programming* problem. If only some of the variables are constrained to take integer values, it is a *mixed integer programming* problem. If the values are confined to take binary values, either zero or one, it is called a *zero-one programming* problem. The integer programming problem could be linear or non-linear. Several methods have been developed for solving the integer programming problem, but the performance of the method depends on the problem or application. The techniques for solving integer programming problems that are either linear or non-linear are the branch and bound method, Cutting Plane method, and Balas method. For non-linear integer programming problems the generalized penalty function method and sequential linear integer programming methods can also be applied. Additionally, a combination of these approaches is also developed [15]. Gomory [16] has proposed an algorithm for obtaining integer solutions to linear programming problems as well as to solve mixed integer programming problems. The *cutting plane method* and *branch and bound method* are efficient in solving linear integer (including mixed integer) programming problems. The branch and bound algorithm was first proposed by Ailsa Land in 1960. The branch and bound method divides the feasible region (solution space) into subregions, and each subregion is further divided (up to as many levels as required) depending on the problem. Branch and bound is more efficient and mostly outperforms the cutting plane method for integer programming problems. It can also be extended to solve mixed integer programming as well as programming problems with binary variables. But the cutting plane method modifies the linear programming solution to obtain the integer solution. It converges quickly in a finite number of steps and was one of the first methods to be developed from which other methods evolved. In addition to the two methods mentioned above, the *Balas* method is also used in solving zero-one integer programming and non-linear integer polynomial programming problem. The non-linear integer programming problems can be classified as general and polynomial. The *general penalty function* method and *sequential linear integer programming* are used for solving general non-linear programming problems.

2.7 Stochastic Programming

Stochastic programming is an optimization method where some of the variables associated with the objective function are random variables with a defined probability distribution. In engineering design problems there could be a minimum and maximum bound for the associated design variables and the actual value might be randomly placed within these bounds. Whenever there are random variables involved in the problem it becomes a stochastic programming problem. Even though the variables associated with the problem are random in nature, the mathematical equations related to the problem make it linear or non-linear, geometric or dynamic, and these problems can be solved using the existing standard techniques.

Example: Minimize the function

$$f(X) = w_1 x_1 + w_2 x_2 + \dots + w_d x_d$$

subject to the constraints,

$$ax_1 - bx_4 > 0$$

$$\sum_{i=1}^{d} x_i = 0$$

where the coefficients w_j, $j = 1, 2, \dots, d$ and the variables a and b are random variables with a uniform probability distribution in the interval [0, 1]. This is a stochastic linear programming problem, and any of the available techniques could be applied to solve them.

2.8 Lagrange Multiplier Method

When the objective function is continuous and differentiable with equality constraints the *Lagrange multiplier* method can be used. Let $f(X) = f(x_1, x_2)$ be the objective function to be minimized or maximized (function of two design variables) and let the constraint be $g(X) = g(x_1, x_2) = 0$. The *Lagrange* function is formulated as:

$$L(x_1, x_2, \lambda) = f(x_1, x_2) + \lambda g(x_1, x_2) \tag{2.4}$$

Let x_1^* *and* x_2^* represent the extreme points on the surface of the function where maxima or minima occur and λ is the Lagrange multiplier. The necessary conditions for x_1^* and x_2^* to be points of maxima or minima are derived from Equation 2.4 by differentiating the function L with respect to x_1, x_2 *and* λ. These conditions are given by Equations 2.5 to 2.7 as detailed below:

$$\left(\frac{\partial f}{\partial x_1} + \lambda \frac{\partial g}{\partial x_1} \right)_{x_1^* x_2^*} = 0 \tag{2.5}$$

$$\left(\frac{\partial f}{\partial x_2} + \lambda \frac{\partial g}{\partial x_2} \right)_{x_1^* x_2^*} = 0 \tag{2.6}$$

$$\left(g(x_1, x_2) \right)_{x_1^* x_2^*} = 0 \tag{2.7}$$

The above three equations are the necessary conditions for x_1^* *and* x_2^* to be the extreme points.

The solution to these equations produces the maximum or minimum value of the function $f(X)$ subject to the constraint $g(X)$. This technique can be extended to problems with multiple constraints by introducing one Lagrange multiplier for each constraint.

2.9 Summary

The traditional optimization methods have been discussed and a few examples have been given wherever possible. The popular traditional algorithms such as linear programming, non-linear programming, quadratic programming, geometric programming, dynamic programming, integer programming, and stochastic programming have been outlined. Some of the optimization techniques under the above categories such as the simplex and revised simplex methods, Kamarkar's method, duality theorem, decomposition principle, and Kuhn–Tucker conditions have been described. The various algorithms are classified based on the number of objective functions, number and type of constraints, number of variables involved, characteristics of the objective function, and so on. The optimization method to be applied depends on the problem and its mathematical equations and most of them are numerical programming methods. The choice of the algorithm depends to a great extent on the dimension of the problem to be solved. Many of the optimization algorithms are iterative with the solutions improving with increasing number of iterations.

The classical engineering design problems such as design of steel and civil structures, optimizing input resources or maximizing output from manufacturing industries, food processing industries, chemical industries, and routing in computer and communication networks are some of the diverse applications of the traditional optimization techniques. The standard transportation problem, duality in problems, and the decomposition principle where large complex problems can be decomposed into smaller units can be solved using different algorithms that are adapted for these types of problems. The sensitivity analysis of an algorithm can be carried out by varying the different parameters associated with the problem and studying the change in the optimal output. The effect of the change in variables, coefficients, constraints, and the number of mathematical equations on the performance of the algorithm and the optimum output can be an area for research. The algorithm should be robust and provide an accurate solution or a solution that is close to the global optimum. If there are any local minima, the algorithm should not get trapped into it and must be able to come out of it. The choice of the algorithm depends on its diversity, versatility, adaptability, flexibility, robustness, and it should be easily programmable to solve a wide array of problems across the entire spectrum of optimization. The space (memory required), time, and computational complexity play a major role in the choice of the algorithm.

References

1. Singaresu S. Rao, *Engineering Optimization, Theory and Practice*, 4th edition, John Wiley & Sons, 2009.
2. Jon C. Nash, The (Dantzig) simplex method for linear programming, *IEEE Computing in Science and Engineering*, Vol. 2, No. 1, pp. 29–31, January/February 2000.
3. N. Kamarkar, A new polynomial-time algorithm for linear programming, *Combinatorica* (Springer), Vol. 4, No. 4, pp. 373–385, December 1984.
4. C. E. Lemke, The dual method of solving the linear programming problem, *Naval Research Logistics* (Wiley), Vol. 1, No. 1, pp. 36–47, 1954.
5. G. B. Dantzig, L. R. Ford D. R. Fulkerson, A primal-dual algorithm for linear programs, In: *Linear Inequalities and Related Systems*, H. W. Kuhn and A. W. Tucker (eds) *Annals of Mathematics Study*, No. 38. Princeton: Princeton University Press, 1956.
6. George B Dantzig, Philip Wolfe, Decomposition principle for linear programs, *Operations Research*, Vol. 8, No. 1, pp. 101–111, February 1960.
7. D. R. Fulkerson, *Hitchcock Transportation Problem, P-890*, Santa Monica, CA: *Rand Corporation*, July 1956.
8. F. L. Hitchcock, The distribution of a product from several sources to numerous localities, *MIT Journal of Mathematics and Physics*, Vol. 20, pp. 224–230, 1941.
9. Philip Wolfe, The simplex method for quadratic programming, *Econometrica*, Vol. 27, No. 3, pp. 382–398, July 1959.
10. R. J. Duffin, E. L. Peterson, and Zener, C., *Geometric Programming*, John Wiley, New York, 1967.
11. Stephen Boyd, Seung-Jean Kim, Lieven Vandenberghe, Arash Hassibi, A tutorial on geometric programming, *Optimization and Engineering* (Springer), Vol. 8, pp. 67–127, March 2007.
12. E. L. Peterson, Geometric programming, In: *Advances in Geometric Programming, Mathematical Concepts and Methods in Science and Engineering*, Vol. 21, M. Avriel (eds). Springer, Boston, MA, pp. 31–94, 1980.
13. H. W. Kuhn and A. W. Tucker, Nonlinear Programming, *Proceedings of the 2nd Berkeley Symposium on Mathematical Statistics and Probability*, University of California Press, Berkeley, 1951, pp. 481–492.
14. Richard Bellman, The theory of dynamic programming, P-550, *Presented to the American Mathematical Society*, Wyoming, July 1954.
15. J. E. Mitchell, Branch-and-cut algorithms for integer programming, In: *Encyclopedia of Optimization*, C. A. Floudas and P. M. Pardalos (eds). Dordrecht, The Netherlands: Kluwer, 2001.
16. R. E. Gomory, Outline of an algorithm for integer solutions to linear programs, *Bulletin of American Mathematical Society*, Vol. 64, No. 5, pp. 275–278, 1958.

3

Nature-Inspired Algorithms

3.1 Introduction

Nature-inspired optimization algorithms are metaheuristic algorithms that are developed from the principles of biological evolution, swarm behavior, and physical and chemical processes [1]. Nature-inspired optimization algorithms are bioinspired computational intelligence techniques since they incorporate intelligence in the algorithms. The research into these algorithms has grown by leaps and bounds in the last two decades. The first breakthrough occurred in the 1960s with the pioneering development of the evolutionary genetic algorithm (GA) by John Holland and his colleagues at the University of Michigan. Since then several evolutionary algorithms have been proposed, including many variants and hybrids of GA. Evolutionary algorithms are based on biological evolution, and GA is one of the classical examples under this category. GP is another popular evolutionary algorithm that is similar to GA and has a population of evolving programs and uses the same operators as GA. Swarm intelligence algorithms are another category of bioinspired algorithms that are inspired by the behavior of swarms in nature such as bird flocking, ant trailing, fish schooling, elephant herding, and so on. Using populations of search agents combined with heuristics has a profound effect on the solutions to complex engineering design problems. The nature-inspired algorithms are novel in attaining effective solutions easily with the least computational resources. The sharing of information and social interaction among members of their own species as well as with the environment by biological agents such as ants, bees, crows, bat, cuckoo, etc. has led to the rise of collective intelligence. They adapt themselves to the environment and make the optimum use of resources available, whether it is sharing of food or any task to be completed, with cooperation amongst their group. Hence any algorithm modeled on their behavior can find solutions to complex problems easily.

The majority of nature-inspired algorithms are broadly classified under evolutionary algorithms (EA) and swarm intelligence (SI) algorithms. The third category of nature-inspired algorithms are based on physical and chemical processes, and simulated annealing (SA) is a famous algorithm under this class. Evolutionary Algorithms are developed from the biological processes of evolution whereas swarm intelligence algorithms are developed from the study of swarm behavior. There has been a breakthrough in the development of nature-inspired algorithms that mimic the behavior of swarms of animals, birds, and insects since the invention of particle swarm optimization (PSO) in 1995. Particle swarm optimization was the first population-based swarm intelligence algorithm to be proposed based on the flocking behavior of birds [2]. The study of nature, flora, fauna, and the ecosystem in general has been the inspiration behind the development of nature-inspired optimization algorithms and hundreds of nature-inspired algorithms,

FIGURE 3.1
Dampa Tiger Reserve Forest. (Author: Coolcolney – own work, CC BY-SA 3.0 https://creativecommons.org/li
censes/by-sa/3.0/deed.en.)

their variants, and hybrids have been proposed. Figure 3.1 shows a picture of Dampa Tiger
Reserve in Mizoram, India, a lush tropical forest that is home to diverse flora and fauna.

3.2 Traditional versus Nature-Inspired Algorithms

Optimization algorithms are a broad class of algorithms with a mathematical foundation that have been designed to find the optimum solution under constraints. Traditional
algorithms do not always guarantee the global optimum solution since the final solution depends on the initial conditions. If they start at the same initial point they arrive
at the same final solution since the traditional algorithms are deterministic. The classical, derivative-based algorithms are problem-dependent and rely on the objective function landscape, so they will not be suitable for problems with discontinuities. Moreover,
they will not be suitable for complex, non-linear, multimodal problems [3]. Any problem
which appears to be extremely complex or hard to solve using traditional methods can be
solved by taking a leaf out of nature. Motivation can be gained by studying nature and
how such problems are dealt with in biological species. Nature-inspired algorithms do
not require computation of derivatives; hence they are gradient-free and are not problem-
specific. Even if the algorithm starts at the same initial point for repeated runs, it will not
end up with the same solution. There is some in-built stochasticity in the algorithm, with
Levy flights and random walks. Since nature-inspired algorithms have started to develop
and show promising results, there has been an explosion in their applications in various
fields. These include engineering, industry, economics, communication, computer science,
networks, business management, etc.

One of the major approaches to engineering optimization is to search among all the feasible solutions, to find the global optimum or the best solution that fits the problem and its constraints. Basically, nature-inspired algorithms are metaheuristic search algorithms that search for the optimum in parallel, with a population of agents [4]. Nature-inspired algorithms generate solutions that are close to the optimum (if not exactly the global optimum) in a finite reasonable amount of time, as opposed to traditional algorithms that are intractable for NP-hard problems. Even if they get stuck in local optima, the in-built randomness enables them to jump out of the local optimum. They can solve linear as well as non-linear problems that are either unimodal or multimodal. Heuristics and metaheuristics incorporate some form of approximation and randomness in the algorithms, have memory to store the past history which could be the best solution attained so far, and they also learn from the past successes. The disadvantage is the large number of iterations required and the lack of consistency in the solutions attained with each iteration. The tradeoff between traditional and metaheuristic algorithms is that the traditional algorithms have good exploitation properties whereas the stochastic algorithms have good exploration capabilities. Hybrid algorithms have been proposed by combining more than one nature-inspired algorithm. This has been found to give better performance than each of the algorithms acting alone. A suitable combination of algorithms is essential to utilize the best properties and characteristics of each.

3.3 Bioinspired Algorithms

Bioinspired computation is a branch of computational intelligence, and the different algorithms in this category are based on the characteristics of biological systems, evolutionary computing, and swarm intelligence [5]. Bioinspired computing has wide-ranging applications in all fields of engineering, especially computer science, economics, mechanical design, and many other real-life application areas. They are more suitable for problems that are computationally complex and data-intensive and found to be intractable to solve using the traditional algorithms. Bioinspired algorithms are efficient in arriving at the optimum solution to a problem when there are myriads of possibilities. They are non-deterministic and are used in analyzing complex systems such as vehicle routing, network routing, job scheduling, and so on. Their simplicity and inherent parallelism are two main reasons for their popularity and wide range of applications [6]. They are flexible and can be made adaptive to the changes in the environment.

Bioinspired algorithms could be trajectory-based or population-based. In trajectory-based algorithms such as simulated annealing the search for the optimum solution starts from a single point initially and gradually reaches the optimum. In population-based algorithms such as genetic algorithm or particle swarm optimization the search for the optimum takes place in parallel by a population of particles or agents in the search space. The search is a tradeoff between wide-area global search (*diversification*) and intense local search (*intensification*). A good balance between these two is essential for finding the global optimum solution in least time [7]. The literature on metaheuristic algorithms has expanded tremendously over the last two decades, and there is lot of scope for development of new algorithms or a hybrid of existing algorithms that shows improved performance over the algorithms acting alone, or variants of the existing algorithms.

Complex engineering design problems often have non-linear constraints such as bounds on certain parameters, relationships among two or three parameters leading to a mathematical representation of a constraint, and the fact that the landscape of the objective function could be unimodal or multimodal. Multiobjective optimization problems often have conflicting objectives and constraints. There is no so-called best solution, but several non-dominated solutions lie on the Pareto optimal front. Feature selection is one of the important problems in image classification where selecting the appropriate features of an object is crucial for classification and hence recognition. The number of features and the appropriateness of the selected features are important in determining the computational cost, memory, and classification accuracy. This is one of the typical problems in optimization for metaheuristic algorithms to solve and prove their efficiency. In general, resources are limited, and with these limitations and possibly some constraints, it is necessary to solve problems to find optimum solutions which could be the maximization of profit or efficiency, or the minimization of time or cost or resources utilized.

Most of the nature-inspired optimization algorithms are heuristic, that is, they find the best approximation which might not be the exact solution to the problem. But these algorithms produce the approximate solution in finite time, by some simplifying assumptions which is not always true for deterministic algorithms. Metaheuristic algorithms work at a higher level than heuristic algorithms and have a tradeoff between direct exhaustive search and randomness. Most of the metaheuristic algorithms incorporate a random parameter with a known probability distribution, to accelerate the search for the optimum solution. There is always some trial and error involved in the search. It is like having solutions lying somewhere in a huge search space that has to be entirely combed in order to find the optimum solution [8]. Since it might not be possible to cover the entire space, we start the search somewhere, assuming it is the right place or assuming the optimum solution lies in a particular region of the search space. This assumption on which the search is based is heuristics, and as the search progresses it will have additional inputs to modify the search or diversify into unexplored regions or intensify the search in a local region. The algorithm should be designed in such a way that the search does not get trapped in any local optimum. When there are multiple agents looking for the solutions in the search space it is population-based, whereas if there is only one agent it is trajectory-based, such as hill climbing or simulated annealing.

3.4 Swarm Intelligence

The study of the self-organizing [9] collective intelligent behavior of swarms has been a topic of extensive research since the mid-1990s. The SI algorithms are very powerful since they have inherent parallelism and are adaptive. The swarm intelligence algorithms are quite effective in solving complex, non-linear optimization problems with reduced time, space, and computational complexity. SI algorithms are characterized by multiple search agents that search for the optimal solution in parallel, thus being efficient. They share information among the members of the swarm, use their collective intelligence, and are self-organized and evolutionary in nature [10]. These algorithms have been designed to solve problems that have been proven to be NP-hard for the classical algorithms. Observing the behavior of a flock of birds flying in nature shows that there is no limit to the number of birds in the flock. The formation of a large flock involves a large number of members

FIGURE 3.2
Flock of birds. (Author: Faisal Akram from Dhaka, Bangladesh, CC BY-SA 2.0 https://creativecommons.org/li
censes/by-sa/2.0/deed.en.)

searching over a large area for quality food and guarding themselves against predators.
Figure 3.2 shows a flock of *Red-billed Queleas* that form enormous flocks that could be tens
of thousands in number.

Flocks of birds usually have a natural formation irrespective of the size of the flock. The
flock is modeled using three rules: collision avoidance, velocity matching, and flock centering. The birds in the flock always maintain a safe distance from their neighbors to avoid
collision and fly with almost equal velocity to remain together. They also maintain their
position with respect to the center of the flock. Figure 3.3 shows a flock of common cranes
(*Grus grus*) flying over Castilla, Spain.

The swarm intelligence algorithms incorporate the foraging strategies of biological
organisms such as animals, birds, and insects. The agent with the best foraging strategy
survives in the environment since food is essential for survival. This involves a search
process, and the agents with the more efficient search strategy succeed quickly compared
to others in the competitive environment. There are several factors involved in such foraging activities – the characteristics and size of the agent, its intelligence, social behavior,
location of food and quantity, and the effort required to find the food. The presence of
predators and other dangers lurking in the environment and the capability of the agents to
ward off such forces and escape from them or protect themselves against such predators
also play a vital role in survival. Moreover, since the environment is dynamic, the quality,
quantity, and location of food keep changing with time due to consumption and other
changes that occur over a period of time. This necessitates the organism to be adaptive
and versatile to the changing conditions. The feedback of the past successes of the flock,
and sharing of information among the flock members regarding food quality and location make it easier to find food. Several species exhibit social foraging which enhances
their chances of success in finding food and hence ultimately leads to better chances of
survival. This also requires good communication skills among members of the population and a sharing strategy. To take care of this, usually there is a hierarchy in the group

FIGURE 3.3
Flock of cranes. (Author: Arturo de Frias Marques – own work, CC BY-SA 3.0 htttps://creativecommons.org/li
censes/by-sa/3.0/deed.en.)

of organisms with a leader who coordinates such activities. Communication of the find
(quality and quantity of food) is done in several ways depending on the species and type
of organism. Some of the members go back to a central location where all the members of
the group are present and disseminate the information about the quality, location, and
distance of the food source. Sometimes, the members are led back to the location of food by
the member which has discovered the food source. In other species, the communication is
done through broadcasting of information which could be heard by members of their own
group as well as by predators, increasing the danger of being attacked. The advantages
and disadvantages of the different types of communication about food discovery among
the different species vary.

In some other species like ants, a visible trail is created between the food source and
the nest or their living place. Typically, creating a chemical trail is done by ants by laying
down pheromones which are followed by other ants, thus increasing the concentration of
the chemical laid on the path as more and more ants follow the trail. This increases the
chances of other insects or predators also finding the food source or attacking the ants.
Broadcasting of a food find is done by crows which invites food calls to other members
of the species. This could be heard not only by members of their own species but also by
other birds and animals, thus increasing the risk of attack. The communication of food discovery by a member by going back to their group or nest is done by honey bees. The bees
perform a waggle dance where the duration of the dance and the orientation of the bee
indicate the quality and location (distance and direction) of the food source. Going back to
the nest to communicate the discovery of food and leading the pack to the food location is
done by some species of birds such as the raven. The last two strategies are advantageous
in terms of safety from predators and eavesdroppers.

The swarm behavior of biological species upon which the SI algorithms are built use
simple rules, and there seems to be no centralized control. When there is no centralized
control, the individual behavior exhibits self-organization and control. Figure 3.4 shows a
flock of Auklets exhibiting swarm behavior with no obvious centralized control.

FIGURE 3.4
Auklet flock Shumagins 1986. [Author: D. Dibenski, U.S. Fish and Wildlife Service (Public Domain).]

Searching a vast area in parallel with many agents and sharing of information collected during the search is the key to success in finding the optimal solution in an effective and efficient manner. The search is started in a random manner with the agents being allotted initial positions randomly. The search also proceeds randomly, and as time passes the information about the findings is shared by the fellow members to either proceed in the same direction or change their area or direction of search. The entire area where there is a possibility of finding a solution has to be searched, and for this good exploration with diversification is necessary. If there is any possibility of finding the solution in a particular area it has to be exploited, and the search has to be intensified around that location. If there are peaks in the region then every peak has to be climbed and searched and the findings shared and compared. If the solution is present at the highest peak then this can be identified only after climbing and comparing the heights of the different peaks. Such simple search strategies can achieve results effectively; this is the reason for their popularity, and a lot of research has gone into the development of such nature-inspired metaheuristic algorithms. The multiple agents in these algorithms interact with each other. This is the fundamental underlying principle behind all such population-based algorithms.

The evolution process in nature has been taking place for millions of years, and new ingenious solutions have been invented in the ever-changing environment. Almost all the species in nature are adaptive and keep evolving to find better ways of solving problems. According to Darwin's theory, it is *survival of the fittest* in all species over the years. Hence the population members have to adapt themselves and think ingeniously for survival. Their success depends on their intelligence and adaptation capability to survive in the ever-changing environment. The pressure of survival in a hostile environment forces the members to think intelligently and adaptively so that they can improve upon the present for the benefit of the next generation.

But many of the species such as elephants and gray wolves have a hierarchy in the herd or pack, and they adhere to the rules laid down by the head of the group. Self-organization requires memory to remember the past (successes or any other matter), communication within the group (or any other interactions), adaptation to the environment, feedback to take corrective action, and sharing of food and information. One important point regarding nature-inspired algorithms is that there is no guarantee that the solution found will be the global optimum. It could be a local optimum, or it could be close to the global optimum if not the exact solution. The landscape of the objective function is used to improve upon the existing candidate solutions so that better solutions evolve as the iterations progress. The new solutions are evaluated on the function landscape, and if a new solution is better than the existing ones, it is included in the population. When new solutions are included, the weaker ones or those with lesser fitness are discarded so that the population size remains constant. Some information on the fitness function landscape can help guide the search so that convergence takes place faster. The diversity of the population must be maintained in order to provide quality solutions and prevent the algorithm from getting stuck in local optima.

Stochastic operators employed in metaheuristic algorithms are responsible for the diversity in the search and faster convergence in large-dimension search spaces [11]. Deterministic algorithms also search the same space, and even if the initial point is same, the stochastic algorithm is faster and does not converge at local optima. Metaheuristic algorithms are more likely to find the global optimum than the classical deterministic algorithms. The main difference and advantage of metaheuristic algorithms over their classical counterparts is the randomness in the search and inclusion of diversity with a population of search agents. The starting point for metaheuristic algorithms is a population of solutions that are initially generated randomly. These solutions are improved with each iteration until a stopping criterion is satisfied or maximum number of iterations is reached or the global optimum solution is attained. Nature-inspired algorithms do not require the computation of derivatives, and they are more efficient than their classical counterparts. The final optimal solution might not be 100% accurate, but a solution quite close to the optimum could be attained. Individual agents could be simple and not very intelligent, but the collective behavior of a population of agents exhibits intelligent and self-organizing behavior able to solve a variety of complex tasks. This collective behavior is the characteristic of flocks of animals, birds, and insects that is the main inspiration behind nature-inspired algorithms. Nature-inspired algorithms are simple to implement and able to tackle non-deterministic NP-hard problems effectively. The nest built by harvester ants using their collective intelligence and cooperation making the best use of available resources is shown in Figure 3.5.

The SI algorithms are sometimes inefficient when the dimensionality of the search space is large. Unless the algorithms have been designed specifically, the search could take a long time over high-dimensional spaces. The search spaces could be structured or unstructured. Using populations of search agents combined with heuristics has a profound effect on the solution of complex engineering design problems. The nature-inspired algorithms are novel in attaining effective solutions easily with the least computational resources. Neighborhood search refers to searching for the optimum solution in the neighborhood of the existing solutions. This will lead to exploration of the variants of the solution. A sequence of steps where the steps are taken in a particular direction (ascent or descent) is equivalent to hill climbing. Some steps could be in the opposite direction in order to enable the algorithm to come out of local optima. Otherwise the search could be started from some other points in the solution space. The history of past successes could be utilized in the current search. In almost all of the evolutionary algorithms, there is a population of search agents that search the solution space, looking for the optimum solution. They are basically iterative in nature, trying to improvise

FIGURE 3.5
Harvester ants' nest. (Author: Indu MG – own work, CC BY-SA 4.0 https://creativecommons.org/licenses/by-sa/4.0/deed.en.)

upon the existing solutions with each succeeding iteration. This improvisation is in the form of a new population which includes the fittest members of the previous generation as well as new members created by some mechanism that is specific to an algorithm. In this process, the weaker members of the population get discarded. The advantage of population-based search is that it takes place in parallel and it could be directed to explore possible regions of the space that were previously unexplored where the optimum solution could be found.

3.5 Metaheuristics

Heuristics is a strategy used when it is not possible to obtain an exact solution in solving a problem in finite time. Applying heuristics gives a satisfactory approximate solution to the problem in practically reasonable time, but it might not be the accurate solution. Using a *rule of thumb* is a problem-solving strategy that is adapted from solving a previously similar problem and is a simple example of heuristics. Heuristics are useful in solving problems that require approximations. Metaheuristics are higher level heuristics [12] used to solve optimization problems, especially those that have incomplete data such as those in artificial intelligence and machine learning. In some of the problems, when the set of solutions is too large to be completely tested, metaheuristics may be applied. Since it is not exhaustive, the global optimum might not be found for all the problems. Metaheuristics might be implemented in stochastic optimization, and in combinatorial optimization it searches over a large discrete set of feasible solutions. They require lesser computations compared to the regular methods. Metaheuristics is a general strategy that is applied to the implementation of a wide range of optimization algorithms [13]. Metaheuristic algorithms always work on any problem and find a solution even though it might not be the best or exact solution to the problem. The highlight is that it arrives at the solution in reasonably finite time suitable for practical applications.

Most of the metaheuristic algorithms are search algorithms, but it is impossible to search every possible candidate solution in the search space so some heuristics are required. Since heuristics are involved in the search, there is no guarantee that the solution will be the best or the global optimum. The solution could be the best one for the problem (globally optimum), or it could be close to the optimum (good approximation). These algorithms converge in finite time in tens or hundreds of iterations, as the case may be. One of the important characteristics of such algorithms is improvisation of the existing solutions with each iteration. The new solutions are better than the existing ones, and the poorer ones are discarded. Following some simple rules of nature, even non-linear problems with constraints can be solved efficiently.

Metaheuristic algorithms are characterized by memory (many of them use past history of successes in the search for the global optimum), collective intelligence, sharing of information, self-organizing, and foraging capabilities of swarms. Following the rules of the swarm with interactions between multiple agents and feedback during the search enables self-adjustment according to the landscape of the objective function. Metaheuristic algorithms have become very popular because of their simplicity and flexibility and the ability to solve NP-hard problems in finite time. They can solve continuous as well as discrete problems and do not require the computation of derivatives. They can easily solve unimodal as well as multimodal problems that have single and multiple optimum respectively. Almost all of the metaheuristic algorithms proposed have been based on nature. It includes biological evolution, behavior of flocks (or herds) of animals, birds, insects, physical and chemical processes, and other mechanisms that occur in nature. The algorithms have been developed from scratch, or they have been built upon an existing algorithm, or two or more algorithms have been hybridized. The striking concept in these algorithms is that they are able to find the optimum solution very quickly and mostly they do not get trapped in local optima. There is always a randomness in the search for the optimum solution by the algorithm. Most of these algorithms are population-based, and they undertake the search in parallel, thus reducing the search time. The stochasticity in these algorithms is the main factor leading to the global optimum without getting trapped in local optima. The associated parameters have to be tuned or adjusted in order to reduce computational and time complexity, thus speeding up the convergence.

The metaheuristic algorithms modeled on this behavior of biological species have been found to be suitable for engineering design, machine learning, artificial intelligence, and a host of other problems which are difficult to solve by traditional methods. There is lot of randomness in nature, and this randomness incorporated into the algorithms helps in improving diversity and jumping out of local optima. Many challenging applications, such as the traveling salesman problem, knapsack problem, feature selection, and image classification, have been solved easily using the metaheuristic computational intelligence techniques. The different members of the population that search in parallel use their collective intelligence and share information which helps in narrowing down the search quickly. Some members who are robust and are good in the search process or find good solutions survive for the next iteration or generation whereas members who are weak and give poor performance are discarded from the search in the next iteration. This is typically Darwin's theory of *survival of the fittest*. This is the crux of all metaheuristic algorithms that are nature-inspired or bioinspired.

3.6 Diversification and Intensification

The main characteristic of metaheuristic algorithms is that the search takes place in two phases: *exploration* and *exploitation*. The exploration capability is searching for the optimum solution in a large search area that has been previously unexplored so as to reach the global optimum. Exploration leads to solution diversity and the ability to jump out of local optima, if any. The exploitation phase is for intensifying the search in a smaller area where it is feasible to find the optimum solution. A good algorithm should properly balance between these two phases optimally for effective and efficient performance [14]. The nature-inspired algorithms are based on survival of the fittest and adaptation to the environment. This leads to two crucial concepts: *diversification* and *intensification*. *Diversification* is the ability to search unexplored areas in the entire search space effectively, whereas *intensification* is exploiting local regions by searching around a current best solution. Some of the techniques could use gradients for such intense local search. The right balance of *exploration* (*diversification*) and *exploitation* (*intensification*) is the key to the success of the metaheuristic nature-inspired optimization algorithms [15].

Some of the species of insects and birds exhibit Levy flight behavior which is straight flight paths punctuated by sharp 90° turns. This Levy flight trajectory can be useful for global exploration of the search space and hence in diversification of the search. Search algorithms mostly use variable step sizes or Levy flights to balance between diversification and intensification. This right balance could be achieved by choosing appropriate values for the parameters associated with the problem. More exploration leads to slower convergence but increases the possibility of finding the global optimum whereas more exploitation leads to faster convergence but the possibility of finding the global optimum is reduced and the probability of the algorithm getting trapped in local optimum is increased. This balance is one of the most important distinguishing factors among the metaheuristic algorithms and is reflected in terms of their performance for various applications. Some algorithms use an intermittent search strategy wherein the fast phase (global explorative phase) and the slow phase (local intense search) are intermittently applied. This intermittent search will be mainly useful for multimodal functions where the search area could be quite large with multiple optima.

The solutions tend to move in the search space. This is possible by the movement of the particles whose evaluation at any point in the objective function landscape is the fitness value or solution at that point. These agents move with each iteration towards regions where they will have a higher fitness value. Thus the average fitness value of the population increases. This is similar to fireflies that are attracted to other fireflies which have higher brightness than themselves. Brownian movement and diffusion of any liquid such as ink, paint, or watercolor on a piece of cloth are equivalent to random movement of particles in exploration. Diffusion is similar to series of finite steps such as Levy flight. The diffusion movement could follow Gaussian or uniform distribution in most of the cases. Other probability distributions are also possible in random walks, but these two are commonly used. In exploitation where the local search is intensified, the randomness could be reduced and the algorithm tends towards deterministic moves. The concept of attraction was studied by the behavior of fireflies and how they congregate at a location because of attraction of their flashing lights. This intensifies the search around the region where the fireflies have congregated and leads to faster convergence of the algorithm. Exploration looks for new solutions in the search space which could possibly be better than the existing solutions and hence the search should be randomized to explore new regions which have been unexplored so far. The step sizes of the random walks could be adjusted between large and small so that newer and better solutions are generated that completely cover the search space. Using the information gained as the search proceeds, the search is intensified in regions around which a possible solution could exist. If there is too much exploration, it takes more time for the algorithm to converge but the probability of finding the global optimum increases. If there is too much exploitation, it takes less time for the algorithm to converge, but the algorithm could converge at a local optimum leading to premature convergence.

3.7 No Free Lunch Theorem

One of the most important concepts in metaheuristics-based nature-inspired algorithms is the *No Free Lunch Theorem*. The *No Free Lunch Theorem* [16] was proposed by David Wolpert and William Macready in 1997 and in effect states that all optimization algorithms are comparable in performance when applied to a wide range of problems across the entire spectrum of engineering optimization. According to the theorem, there is no optimization algorithm that is better than any other algorithm; each one is best-suited for a particular class of optimization problems. The better performance of an optimization algorithm over one class of problems is offset by the performance over another class of problems. An optimization algorithm that gives an optimum result for one problem might not produce an optimum solution when applied to another problem, whereas another optimization algorithm might produce the optimum result for the second problem. To put it in another way, each algorithm is best-suited for a particular class of problems, but it might not perform equally well for all types of problems. In essence, the theorem states that the various optimization algorithms cannot be ranked and compared based on their performance with one class of problems. The algorithms cannot be categorized as good or bad, and there is no best or worst algorithm as such. The algorithms are applicable to all disciplines like aerospace, automobile, electrical, electronics, communication, civil, instrumentation, chemical, textile, computer science, economics, management, etc. The *No Free Lunch Theorem* states that all optimization algorithms are, in effect, equal [17]. Rapid advances have taken place in the development of nature-inspired metaheuristic algorithms with each one suitable for a particular type of application, thus enforcing the *No Free Lunch Theorem*.

3.8 Parameter Tuning and Control

The performance of nature-inspired optimization algorithms depends on the setting of the parameters associated with the problem. Choosing the appropriate values for the parameters initially and maintaining them throughout the run of the algorithm is parameter tuning. As the number of iterations increases, the parameters could be kept constant or they could be varied adaptively. If the parameter values are modified as the iterations progress, it is parameter control. In some applications, varying the parameters as the algorithm progresses leads to faster convergence and attaining of global optimum. This is referred to as parameter tuning and control [18].

In most of the algorithms in the literature, the parameters have been tuned based on experimental observations and results. The algorithms have been run on several different test problems and standard benchmark functions and experimentally the parameters have been tuned. The balance between *intensification* and *diversification*, faster convergence, lesser computational time, jumping out of local optima, and approaching the global optimum can be fine-tuned with parameter tuning and control. This leads to better performance of the algorithm which could be in terms of convergence time or the accuracy of the solution attained. When parameter control is done, it automatically takes care of the tuning problem. The initial values of parameters could be chosen randomly, and as the iterations progress, according to the landscape of the search space, the parameters could be changed dynamically. The feedback about the performance of the algorithm, probably in terms of fitness function values, could be used for parameter control. Population size is one of the important parameters for algorithm performance in terms of search complexity, interaction, social sharing of information, and rate of convergence. The fitness function could also be altered dynamically in certain algorithms. One typical example is multiobjective optimization, where the objective (fitness) function is formulated as a weighted combination of multiple objectives, wherein the weights could be altered as the iterations progress. In some of the constrained optimization problems, the equations for the constraints usually have some constants or weights or bounding values. These weights could also be possibly altered dynamically.

Parameter tuning and control are algorithm-specific, and a tradeoff could be necessary. Some algorithms could require parameter tuning alone while there may be others that require parameter control also for enhanced performance. A swarm intelligence algorithm could be tested with the traditional method of having constant parameters, with parameter tuning alone, and with parameter control also included. This could serve as a topic for further research in this area. The variation in the objective function value with respect to the parameters of the algorithm is sensitivity analysis of the optimization algorithm. Sensitivity analysis of the algorithms will help in improving the performance of the algorithms and the spectrum of applications.

3.9 Algorithm

Let S represent the set of possible solutions for $f(X)$ and for each member of S (one possible solution to the problem) there is a neighborhood. We start with an initial subset of solutions from the set S and search in their neighborhood iteratively until the feasible solution is found. This is the local search. If the feasible or optimum solution is not found in the neighborhood of the chosen initial solutions, the search expands over the space. This is the global search. The iterations continue until a stopping criteria is reached or maximum number of iterations is attained. With each iteration the search continues in a direction such that the cost is minimized. If the cost is smaller than that of the previous iteration, the new solutions replace the old ones; otherwise it is discarded and the search continues. Since the search is initially in the local neighborhood of the subset of solutions chosen, there must be some means to make the algorithm come out of local optima and diversify the search. The final solution attained could be the global optimum or local optimum, depending on the effectiveness of the algorithm.

The following properties are preferable in any optimization algorithm:

- The objective function could be linear or non-linear, differentiable or non-differentiable, simple or complex, maximization (quality/fitness) or minimization (cost/error) function.
- The constraints could be linear or non-linear, simple or complex, equality or inequality, and they could be one or multiple.
- The number of parameters should be lower, and they must be easily tunable or adjustable to get good performance.
- The dimension of the search space should be manageable by the algorithm with the given population size.
- The algorithm should have a faster rate of convergence, i.e. it must converge in a smaller number of iterations.
- The algorithm should have less time, space, and computational complexity.
- The algorithm should be self-organizing, and it should have a termination criteria other than the maximum number of iterations.
- Some randomness or stochasticity should be incorporated in the algorithm in order to diversify the search and come out of local optima, if any. The random parameters follow a probability distribution that is usually uniform or Gaussian, although other distributions can also be used.

3.10 Pseudocode

Initialization

Population size N

Objective Function $f(X)$

Constraints $g(X)$ and $h(X)$

Randomly position the members of the population in the search space

Define stopping criteria, if any

Maximum number of iterations *MaxIter*

iter = 1

while (*iter* $\leq$ *MaxIter*) **do**

Execute the algorithm on the population N

Evaluate the fitness of the population

Choose the fittest N members and discard the weaker ones

if stopping criteria met, exit, **otherwise** continue

iter = *iter* + 1

end while

Highest fitness value is the global optimum solution

Flowchart

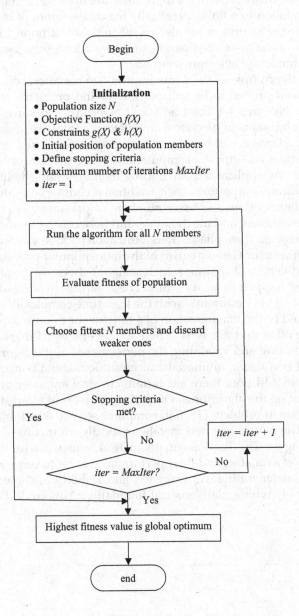

```
                    ┌──────────┐
                    │  Begin   │
                    └──────────┘
                         │
                         ▼
          Initialization
          • Population size N
          • Objective Function f(X)
          • Constraints g(X) & h(X)
          • Initial position of population members
          • Define stopping criteria
          • Maximum number of iterations MaxIter
          • iter = 1
                         │
                         ▼
          Run the algorithm for all N members
                         │
                         ▼
          Evaluate fitness of population
                         │
                         ▼
          Choose fittest N members and discard
          weaker ones
                         │
                         ▼
              Stopping criteria met?
            Yes ◄──          ── No
                         │
                         ▼
              iter = MaxIter?        ── No ──► iter = iter + 1
                         │
                        Yes
                         ▼
          Highest fitness value is global optimum
                         │
                         ▼
                    ┌──────────┐
                    │   end    │
                    └──────────┘
```

3.11 Summary

Nature-inspired algorithms are metaheuristic and are found to be able to solve challenging problems in optimization efficiently. The behavior of the gray wolf, cuckoo, crow, firefly, bee, ants, and, in general, any swarm of particles has been studied extensively in the literature, and their behavior has been put to use to solve practical real-life problems. Many of the real-world applications are highly non-linear, requiring state-of-the-art optimization techniques for solving them. Most of the algorithms are iterative, and they converge to the optimum or final solution in a finite, practically reasonable amount of time. The solution depends not only on the algorithm but also on the initial conditions. Choosing the appropriate initial values of the associated parameters goes a long way towards faster convergence of the algorithm to the optimum solution.

Nature-inspired algorithms, as the name implies, are based on some natural phenomenon. They have been proven to be suitable for solving problems efficiently that have been intractable or NP-hard for the traditional algorithms to solve. The optimization problem could be continuous or discrete (combinatorial) in nature. Nature-inspired algorithms are flexible, adaptive, self-organized, and population-based with simple interactions among individuals and efficient computations. More variants of existing algorithms, improvements, and new applications are finding their way into the literature every day. Metaheuristic algorithms outperform their traditional counterparts due to the search in parallel by a population of agents, lesser numbers of parameters to tune, the ease and simplicity of implementation, and the dynamic shift between exploration and exploitation (*diversification* and *intensification*) phases. Randomization by following some stochastic distribution such as uniform or Gaussian (two of the more popular probability distributions) helps in achieving diversity of solutions. Increasing the diversity of solutions reduces the possibility of getting trapped in local optima. These nature-inspired algorithms can deal with complex problems very efficiently with the least time complexity.

The problems could be the maximization of a function or minimization since one could be converted to the other and *vice versa*. There are several benchmark functions that are used as test set for testing and validating the performance of the algorithms. The test set includes linear and non-linear, unimodal and multimodal, and constrained and unconstrained functions. In addition, there are certain classical engineering design problems that are used for testing the algorithms. Of all the problems of interest, the most famous is the traveling salesman problem (TSP) in computer science which has been researched for several years. The nature-inspired metaheuristic algorithms have been found to be promising for solving the TSP. Rigorous mathematical framework for these metaheuristic algorithms are not yet available, and their performance is yet to be proven for large-scale problems. The parameter tuning, convergence, and stability of these algorithms are to be proven theoretically with a mathematical foundation. This could be a fertile area for further research.

References

1. Iztok Fister Jr., Xin-She Yang, Iztok Fister, Janez Brest, Dusan Fister, A brief review of nature-inspired algorithms for optimization, *Elektrotehniški Vestnik*, Vol. 80, No. 3, pp. 1–7, July 2013.
2. Petr Bujok, Josef Tvrdik, Radka Polakova, Comparison of nature-inspired population-based algorithms on continuous optimization problems, *Swarm and Evolutionary Computation* (Elsevier), Vol. 50, Article ID 100490, November 2019.
3. Xin-She Yang (Ed.), Nature inspired algorithms and applied optimization, *Studies in Computational Intelligence* (Springer), 2018.
4. Xin-She Yang, Nature inspired metaheuristic algorithms: Success and new challenges, *Journal of Computer Engineering and Information Technology*, Vol. 1, No. 1, pp. 1–3, November 2012.

5. Zhihua Cui, Rajan Alex, Rajendra Akerkar, Xin-She Yang, Recent advances on bioinspired computation, *The Scientific World Journal*, Hindawi Publishing Corporation, Vol. 2014, Article ID 934890, May 2014.

6. Scott McCaulay, Biologically inspired computing algorithms: Relevance and implications for research technologies, Indiana University, Bloomington, IN. PTI Technical Report PTI-TR12-003, February 2012.

7. Xin-She Yang, *Nature-Inspired Metaheuristic Algorithms*, 2nd edition, Luniver Press, 2010.

8. Xin-She Yang, *Nature Inspired Optimization Algorithms*, 1st edition, Elsevier, London, 2014.

9. Xin-She Yang, Swarm intelligence based algorithms: A critical analysis, *Evolutionary Intelligence* (Springer), Vol. 7, No. 1, pp. 17–28, April 2014.

10. Xin-She Yang, Suash Deb, Simon Fong, Xingshi He, Yu-Xin Zhao, From swarm intelligence to metaheuristics: Nature inspired optimization algorithms, *IEEE Computer*, Vol. 49, No. 9, pp. 52–59, September 2016.

11. A. Hanif Halim, I. Ismail, Bio-inspired optimization method: A review, *NNGT Journal: International Journal of Information Systems*, Vol. 1, pp. 12–17, July 2014.

12. Xin-She Yang, Su Fong Chien, Tiew On Ting, Computational intelligence and metaheuristic algorithms with applications, *The Scientific World Journal*, Hindawi Publishing Corporation, Vol. 2014, Article ID 425853, December 2014.

13. Michael A. Lones, Metaheuristics in nature inspired algorithms, *Proceedings of Genetic and Evolutionary Computation Conference (GECCO Comp '14)*, Vancouver, BC, Canada, pp. 1419–1422, July 2014.

14. Xin-She Yang, Suash Deb, Simon Fong, Metaheuristic algorithms: Optimal balance of intensification and diversification, *Applied Mathematics and Information Sciences, An International Journal*, Vol. 8, No. 3, pp. 977–983, May 2014.

15. Xin-She Yang, Suash Deb, Thomas Hanne, Xingshi He, Attraction and diffusion in nature-inspired optimization algorithms, *Neural Computing and Applications*, Vol. 31, No. 7, pp. 1987–1994, July 2019.

16. D. H. Wolpert, W. G. Macready, No free lunch theorems for optimization, *IEEE Transactions on Evolutionary Computation*, Vol. 1, No. 1, pp. 67–82, April 1997.

17. C. Schumacher, M. D. Vose, L. D. Whitley, The no free lunch and problem description length, *Proceedings of the 3rd Annual Conference on Genetic and Evolutionary Computation (GECCO '01)*, San Francisco, CA, United States, pp. 565–570, July 2001.

18. Giorgos Karafotias, Mark Hoogendoorn, A. E. Eiben, Parameter control in evolutionary algorithms: Trends and challenges, *IEEE Transactions on Evolutionary Computation*, Vol. 19, No. 2, pp. 167–187, April 2015.

4

Genetic Algorithm

4.1 Introduction

The genetic algorithm (GA) belongs to the class of evolutionary optimization algorithms that is based on the fundamental Darwinian theory of evolution and biological reproduction. It incorporates the principles of *natural selection* and *survival of the fittest*. GA was developed by John Holland, Professor of Electrical Engineering and Computer Science at the University of Michigan, and his colleagues in 1960 followed by a path-breaking publication *Adaptation in Natural and Artificial Systems*, 1975, MIT Press [1]. It was later on extended and developed by David E. Goldberg, and his book on genetic algorithms was published in 1989 [2]. GA emulates the techniques of evolution that have been in existence for millions of years in nature. When an algorithm is designed based on natural genetics it will be able to solve complex problems with simple techniques in finite time.

The study of the biological evolution of humans has led to a deeper understanding of the natural evolution process and *survival of the fittest* strategy, and these techniques or principles have been applied to artificial systems that mimic natural evolution. Natural systems are robust and efficient, and they have the capacity to adapt themselves to the environment which is difficult to emulate in artificial systems. The inculcation of biological evolution principles into artificial systems has led to the development of GA whose performance has been proven and validated for a host of complex problems over the years since its inception. GA uses heuristics as well as the history of the problem in evolving towards newer and better solutions. It tries to mimic human behavior in evolution and search for optimum solutions. There is an exchange of genetic material (information) in the form of strings that are encoded based on the problem [3]. It has found wide-ranging applications in engineering design, business, finance, and other complex scientific fields due to its simplicity and low computational complexity.

4.2 Basics of Genetic Algorithm

The breakthrough in developing GA by John Holland was in the 1960s when he was able to develop codes to represent genetic information. This earliest development was with strings of binary bits where each bit could possibly represent a characteristic or feature. If the characteristic is present, the corresponding bit is 1, and if it is not present, the bit is 0. The system was developed initially as a classifier, where the result or accuracy of classification depended on the value of the string (fitness value) that encodes the features. Strings

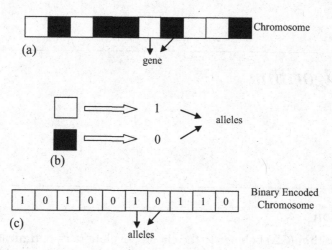

FIGURE 4.1
(a) Typical chromosome structure. (b) *Allele* – value assigned to *gene*. (c) Binary encoded *chromosome*.

with high fitness values yield good results and survive, whereas strings with low fitness values perish. The strings evolve with every generation, leading to a population of strings with higher average fitness (quality) values [4].

The basic genetic material is the *chromosome* that is composed of *genes*. The fundamental element of the *chromosome* is the *gene*, and the value taken by the gene is the *allele*. The *alleles* could be binary with an alphabet of {0, 1} or non-binary such as the decimal alphabet {0, 1, …, 9} or hexadecimal alphabet {0, 1, 2, …, 9, A, B, C, D, E, F} or any other set of defined alphabets. A typical *chromosome* structure, *alleles*, and encoded *chromosome* are shown in Figure 4.1a–c.

Reproduction by mating ensures that the next-generation population or offspring are different from their parents instead of being photocopies of their parents. The genetic material of two members of the population who mate together is fused or crossed over to produce the offspring. Thus the offspring inherits the traits of both the parents. The basic principle underlying evolution is *survival of the fittest*. This ensures that only fit individuals are *selected* for mating and reproduction and the weaker ones perish naturally. The mating process that takes place biologically is done by the operation of *crossover* in GA. Two strings (members of the population with high fitness values) are chosen for mating. Randomly a point is chosen on the string, called the *crossover* point. The bits in the first string after the crossover point are exchanged with the corresponding bits in the second string. This creates two new offspring in the population. These two offspring (children) will replace two members of the population who have low fitness values, thus keeping the population size constant. The next important operator in GA is *mutation*. Some of the bits in the string are randomly chosen and flipped (1 to 0 or 0 to 1). The percentage of strings undergoing *mutation* is very small, typically less than 1%. This ensures some diversity in the population and the production of offspring with new characteristics. *Mutation* does not require parents or mating and leads to change in the genetic characteristics of a population. The change in the genetic code of the parent is embedded in the offspring. The best or optimum solution to the problem will be in the form of a binary string whose fitness value is the optimum.

The search space can be modeled by the population of strings (fitness values) and assuming that the objective function is to be maximized, the peaks correspond to higher quality solutions, and valleys correspond to poor quality solutions. When the number of variables increases, i.e. dimensionality of the problem increases, the landscape becomes complex. In the landscape of solutions, GA exploits the regions with higher fitness values. Strings that have matching bits (either 0 or 1) in certain positions define regions in the search space. For example, 1100 0011 and 1101 0101 match in the first three bits and all strings with their first three bits as '110' form a region in the search space. The *crossover* and *mutation* operators produce the next-generation population with higher average fitness (replacing the older generation members that have lower fitness). Therefore the new-generation members will be located in the higher fitness regions of the landscape, and the population in these regions increases with every consecutive generation in GA. One string will appear in or belong to several regions in the landscape. This makes sampling of several regions occur simultaneously in GA, leading to implicit parallelism [5].

Crossover creates new strings that could possibly belong to a new region different from that of its parents. The distance between 1s and 0s determines the probability of an offspring leaving the region of its parents. For example, let there be a parent string 110***** that crosses over with another string at the third position. One offspring will have the same set of bits in the first three positions, thus belonging to the same region of the landscape as that of its parent. In another case, if the string is 00****11 and the crossover position is anywhere between two and six in the string, the offspring will move to a region different from that of its parent. The 1s and 0s of the region form *building blocks*, and if they are close together (like 110*****), it is a *compact building block*, unlike strings such as 10****11. Regions with *compact building blocks* mostly produce offspring that belong to the same region. This *building block hypothesis* is discussed in detail and mathematically formulated in a later section of the chapter. Another operation called *inversion* in GA rearranges strings in the parents so that bits that are far apart in the parents come close together in the offspring. This *inversion* helps in building compact blocks so that *crossover* will not displace the offspring. The more compact building blocks ensure higher fitness individuals are produced and automatically lower fitness individuals from less compact building blocks will get displaced or perish. When two strings with two different building blocks are combined together and if the combined fitness value is greater than that of their individual fitness put together (added), it implies non-linearity. Hence, GA is also able to solve non-linear problems. In linear problems, the presence of 1 or 0 in a position of the string has no effect on 1 or 0 in any other position of the string. Therefore, a 4-bit binary string will have to explore only eight possibilities. If it is non-linear, the bits at every position will affect the bits in other positions, and the number of possibilities to be explored increases exponentially.

4.3 Genetic Operators

GA has three important operators in the design of the algorithm. They are *selection, crossover*, and *mutation* [6]. These three operators are discussed below:

Selection is an operation in GA for selecting members of the population for mating and recombination/reproduction. The selection process has randomness built into it because there is no deterministic rule applied for the selection of *chromosomes* to carry out *crossover* or *mutation*. The three commonly used methods for *selection*

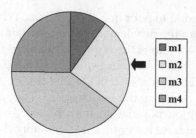

FIGURE 4.2
Roulette wheel.

are roulette wheel selection, tournament selection, and elitism [7]. Roulette wheel selection is one of the important and commonly used strategies in GA for selection of parents in reproduction. A roulette wheel is a rotating wheel divided into various regions (similar to a pie chart), with each region corresponding to one member of the population [8]. The size of each region is proportional to the probability of choosing the member, and this in turn depends on the fitness of the member of the population. There is a pointer associated with the wheel, and when the wheel is spun, the region of the wheel to which the pointer points when the wheel stops spinning is the chosen member of the population. A typical roulette wheel is shown in Figure 4.2.

The roulette wheel is biased according to the fitness value of each string (member) in the population. In the roulette wheel shown in Figure 4.2, the portion corresponding to member m_1 is 10%, m_2 is 25%, m_3 is 40%, and m_4 is 25%. This division is proportional to the fitness values of the four strings in the population. When the wheel is spun, each string is selected according to its probability, such as 0.4 for string 3 (m_3). Strings with higher fitness values have a higher probability of getting selected; hence they get selected for reproduction a greater number of times compared to other strings with lesser fitness values and thus contribute more to the next-generation population. By spinning the roulette wheel several times, a mating pool is created from which members are selected randomly for *crossover* and *mutation*.

In *tournament selection*, randomly some members are chosen from the population (*tournament* size) and from these chosen few, the members with the highest fitness become parents. This can be repeated until the required number of offspring are created. *Tournament* size depends on the size of the population. In *elitism*, the chromosome with the highest fitness value is copied into the next generation so that it will not get modified during *crossover* or *mutation*. The members with fitness values that are not so high (not in the highest range of individuals) are selected for reproduction. This ensures that the highly fit individuals are retained in the population so that the average fitness of the forthcoming generations are higher and the fit individuals are propagated.

Crossover is an operator that produces two offspring from two parents selected for reproduction/recombination. *Crossover* may be one-point, two-point, or multipoint. In *one-point crossover* shown in Figure 4.3a and b, one bit of the two chromosomes (parents) is randomly selected and the two strings are interchanged at the point of crossover, producing two offspring.

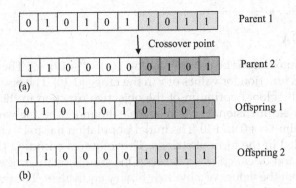

FIGURE 4.3
(a) Single-point crossover (parents). (b) Single-point crossover (offspring).

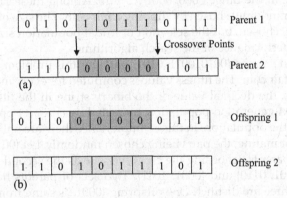

FIGURE 4.4
(a) Two-point crossover (parents). (b) Two-point crossover (offspring).

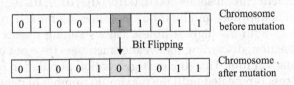

FIGURE 4.5
Mutation.

In *two-point crossover* shown in Figure 4.4a and b, two points are selected in the two chromosomes (parents) randomly and one segment is interchanged between the two parents, producing two offspring.

Mutation is the operator where one or more bits in the chromosome string are flipped with a fixed probability defined as *mutation rate*. *Mutation* is illustrated in Figure 4.5. This introduces diversity in the population and tends to extend the search space effectively. *Mutation rate* is usually kept low in order to maintain the good characteristics of the population with every generation. A high *mutation rate* is not usually desirable since it disrupts the existing good characteristics of the population. *Mutation rate* is usually chosen as inverse of length of the chromosome string, which means that for a chromosome length of 10, the *mutation rate* is 0.1. *Mutation rate* can be changed, either increased or decreased, as the iterations progress.

4.4 Example of GA

Let a function to be maximized be represented by $f(x)=x^2+10x-2$. The problem is to find the maximum of this function for values of x in the range [0, 10]. The optimization problem consists of finding the global optimum of the objective function in the one-dimensional search space. The one-dimensional search space is bounded by the variable x taking on values between the limits of 0 and 10. The initial population has to be chosen in the range of 0 to 10 and encoded in the form of strings. This requires encoding the parameter x as binary strings of 1s and 0s called *chromosomes*. One simple method of encoding the *chromosomes* is to represent the values of x by their binary equivalent. The number of bits used to represent x is the length of the *chromosome*. Since encoding decimal digits up to a maximum value of 10 requires a minimum of 4 bits, the *chromosome* length could be chosen as 4. The strings will be in the range 0000, 0001 ..., 1010. Among these 11 strings, the initial population could be randomly chosen by flipping a fair coin. The size of the initial population is also randomly chosen, but the selection of initial population size and its members has an effect on the performance of the search algorithm.

Let the initial population be {0010, 0011, 0101, 1001} for this problem. This could be chosen with 16 flips of a fair coin. The fitness value is computed for every member of the population by substituting the decimal value of the binary string in the fitness function, thus yielding $f(x)$. The next-generation population is created by *crossover* operation among the selected members of the population. In this example, since there are only four members, all of them are chosen for mating, the pairs being chosen randomly. Let {0010, 0101} and {0011, 1001} be the chosen parents. Choosing the crossover site as the second place in the string, the offspring are {0001, 0110} and {0001, 1011}. Two sets of parents have produced four offspring, of which three are distinct. One offspring {0001} is same from both the parents, so only one copy of this is included in the population pool. Another offspring {1011} is out of bounds of the search space so it is discarded from the population pool. Now we have a population of six different chromosomes {0001, 0010, 0011, 0101, 0110, 1001} or individuals (including parents and offspring), and their fitness values are ranked in descending order. The four chromosomes with the highest fitness values among the six are chosen as the next-generation population, discarding the weaker members (two out of six), keeping the population size constant at four. Thus the next-generation population is {0011, 0101, 0110, 1001}. The above process is repeated until the maximum number of iterations is reached or a stopping criterion is met.

Another variation that could possibly be introduced is the *mutation* operator, in addition to *crossover*. In *mutation*, some strings are randomly chosen and certain bits are flipped, again randomly. The percentage of *mutation* operations compared to *crossover* is very low, typically less than 1%. The fitness of the strings is computed by evaluating the objective function $f(x)$ with the string values. The maximum value of $f(x)$ attained at the end of the iterations is the optimum value of the function and the optimal solution to the problem. The value of x at which this optimum is attained is given by the decimal equivalent of the binary string that encodes the *chromosome* with highest fitness. Table 4.1 shows the fitness function values for the initial population in the example given and their percentage of the total fitness of the entire population. The probability of a string being selected for reproduction is directly proportional to this percentage.

The roulette wheel that has been proportioned for the four members of the initial population in the above example is shown in Figure 4.6. By spinning the roulette wheel several times, the parents for the mating pool are selected.

TABLE 4.1

Fitness Values of the Initial Population

No.	String	Value of x	$f(x) = x^2 + 10x - 2$	% of Total
m_1	0010	2	22	7.31
m_2	0011	3	37	12.29
m_3	0101	5	73	24.25
m_4	1001	9	169	56.15
Total			301	100

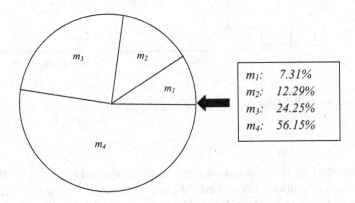

FIGURE 4.6
Roulette wheel selection for Table 4.1.

The operations of *selection*, *crossover*, and *mutation* are repeatedly applied to the population until the maximum number of iterations is reached or the termination criterion is satisfied.

4.5 Algorithm

GA is a population-based metaheuristic algorithm that incorporates natural selection and genetics. GA searches for the optimum solution in the search space with inclusion of randomness in the search. It is an iterative algorithm that converges in a finite number of iterations, with the candidate solutions evolving towards higher fitness or quality with each iteration or generation. The algorithm terminates when either the optimum solution is attained (convergence) or when the maximum number of iterations (generations) is completed or a stopping condition is reached. The important characteristics of GA include initial population selection, defining an objective or fitness function, genetic operators, and termination criteria.

The algorithm starts by defining the problem and its objective function $f(X)$ where X is a multidimensional vector, with a typical dimension of d. The initial population is chosen randomly in the search space, and the members are encoded as a chromosome in the form of a string of alphabets. The operations of *selection*, *crossover*, and *mutation* (optional)

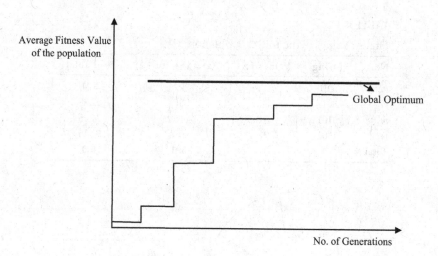

FIGURE 4.7
Typical convergence characteristics of GA.

are repeatedly applied on the population until the termination criterion is attained or the maximum number of iterations is reached. At the end of every iteration, the fitness values of the population of strings are calculated. The strings with higher fitness are selected for mating and reproduction. Finally, the string with the highest fitness value is the optimum solution to the problem.

Typical convergence characteristics of GA are shown in Figure 4.7.

4.6 Pseudocode

Initialization

 Select initial population of size N
 Define objective function $f(X)$
 Encode the population as chromosomes (bit strings) of length L_C
 Compute fitness values of the entire population
 Define termination condition, if any
 Choose maximum number of iterations *MaxIter*
 iter = 1
while (*iter* ≤ *MaxIter*)
 Selection: select parents for reproduction
 Crossover: apply crossover on parents to produce offsprings
 Mutation: apply mutation on selected chromosomes (*optional*)
 Compute the fitness values of the population

Select members for the next generation based on fitness values

If termination condition met exit, **else** continue

iter = iter + 1

end while

Chromosome with highest fitness is the global optimum solution

---

Flowchart

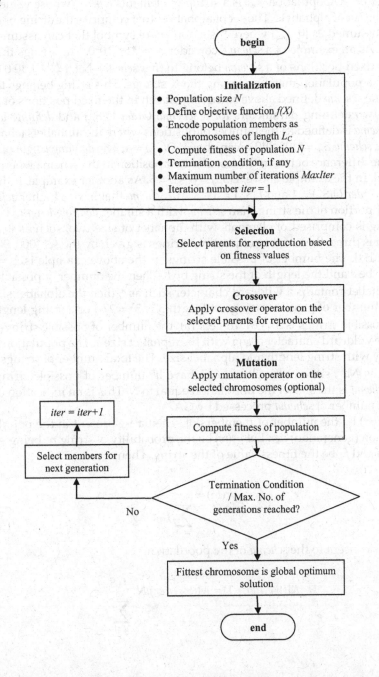

4.7 Schema Theory

Let the population of strings be represented by P^{iter} where *iter* is the iteration or generation number. The population size is chosen as N.

$$P^{iter} = \{p_1^{iter} \; p_2^{iter} \; .... \; p_N^{iter}\} \tag{4.1}$$

where p_m^{iter} is the *mth* member of the population P^{iter} in iteration *iter* and $m = \{1, 2, ..., N\}$. Each member of the population p_m^{iter} is a string of elements (*alleles*) whose values are taken from the given set of alphabets. The set of alphabets that comprise the string or *chromosome* or *schema* is assumed as {0, 1, *} where * is a don't-care symbol that can assume a value of either 0 or 1. As an example of a *schema*, consider $S_{ch} = $ *10* *0*0. The strings that match in alleles at the fixed positions of a *schema* belong to that *schema*. Let $p_m^{iter} = 1\,1\,0\,0\,0\,0\,1\,0$ be a string that is a population member among the N strings. This string belongs to and is an example of the *schema* defined above since they match in the fixed positions of 2, 3, 6, 8.

There are two defining parameters for a *schema*: *Order* $O(S_{ch})$ and *defining length* $L(S_{ch})$. *Order* of a *schema* is defined as the number of positions where the alphabets (characters) are fixed. For the *schema* $S_{ch} = $ *10* *0*0 given above, $O(S_{ch}) = 4$. The *defining length* of a *schema* is defined as the difference between the last and first position in the schema where the alphabets are fixed. In the example above, $L(S_{ch}) = 8 - 2 = 6$. As another example, if the *schema* is $S_{ch} = $ **1* ****, *order* $O(S_{ch}) = 1$ and $L(S_{ch}) = 3 - 3 = 0$. *Schema* that have a higher *defining length* span a larger portion of the string than *schema* with a smaller *defining length*.

If the string is comprised of elements with alphabet of size two, such as {0, 1} and the string length is three, the number of possible strings is given by the set {000, 001, 010, 011, 100, 101, 110, 111}. The number of possible strings in the above example is $2^3 = 8$. Let the alphabet size be a and the length of the string be l_s. Then the number of possible strings is a^{l_s}. If the alphabet contains a wildcard character such as *, then the alphabet size is $(a + 1) = 3$, and the number of possible strings of length 3 is $3^3 = 27$. For a string length of 4, the number of possible strings is $3^4 = 81$. In general, the number of possible strings is $(a+1)^{l_s}$ including the wildcard character along with the alphabet size a. The population size grows exponentially with string length and alphabet size. The total number of strings processed by a schema is Na^{l_s}, since each schema can have a^{l_s} number of possible strings and the number of *schema* is the size of the population equal to N. This is an indication of the magnitude of the number of *schema* processed by GA.

Let $M(S_{ch}, iter)$ be the number of members of schema S_{ch} present in the population during the iteration (generation) *iter*. Let $pr(m)$ be the probability of string p_m being selected for reproduction and f_m be the fitness value of the string. Then,

$$pr(m) = \frac{f_m}{\sum\limits_{m=1}^{N} f_m} \tag{4.2}$$

Applying this concept to the *schema* of the population,

$$M(S_{ch}, iter + 1) = M(S_{ch}, iter).N.\frac{f(S_{ch})}{\sum\limits_{m=1}^{N} f_m} \tag{4.3}$$

$$M(S_{ch}, iter + 1) = M(S_{ch}, iter) . \frac{f(S_{ch})}{f_{av}} \tag{4.4}$$

$$f_{av} = \frac{\sum_{m=1}^{N} f_m}{N} \tag{4.5}$$

$f(S_{ch})$: average fitness of schema S_{ch}

f_{av} : average fitness of entire population of size N

The members of the *schema* grow at the rate proportional to the ratio of average fitness of the *schema* to the average fitness of the entire population. *Schema* with average fitness value greater than the population average grow, whereas *schema* with average fitness less than the population average decay. Therefore, every *schema* grows or decays according to its average fitness value compared to the average fitness of the whole population.

Let $f(S_{ch}) = f_{av} + r.f_{av}$ where r is a positive constant. Then

$$M(S_{ch}, iter + 1) = M(S_{ch}, iter) . \left(\frac{f_{av} + r.f_{av}}{f_{av}} \right)$$

$$= M(S_{ch}, iter) . (1 + r) \tag{4.6}$$

$$= M(S_{ch}, 0) . (1 + r)^{iter}$$

The *schema* with above average fitness contribute exponentially increasing members to the population, whereas there is a decline in the members belonging to the *schema* with below average fitness.

Let p_m^{iter} be the *mth* member of the population P^{iter} and assume $p_m^{iter} = 10101000$. Let $S_{ch}(m_1)$ and $S_{ch}(m_2)$ be two *schema* representative of the string $p_m^{iter} = 10101000$ with $S_{ch}(m_1) = $ *01***** and $S_{ch}(m_2) = $ **1****0. Let the crossover site be chosen between positions 4 and 5. *Schema* $S_{ch}(m_1)$ will survive after the crossover since the fixed positions in the string are on one side of the crossover point, whereas the schema $S_{ch}(m_2)$ will not survive since the fixed positions are on either side of the crossover point. This implies that *schema* with shorter *defining length* survive crossover operation whereas *schema* with longer *defining length* have lesser chances of survival after crossover.

The *defining lengths* for the two chosen *schema* are:

$$L\left(S_{ch}(m_1)\right) = 3 - 2 = 1 \text{ and } L\left(S_{ch}(m_2)\right) = 8 - 3 = 5$$

The probability of a *schema* being destroyed is

$$p_d\left(S_{ch}\right) = L\left(S_{ch}\right) / \text{No. of crossover points}$$

No. of crossover points = length of the string $- 1 = l_s - 1$

The probability of the two *schema* being destroyed is

$$p_d\left(S_{ch}(m_1)\right) = 1/7 \text{ and } p_d\left(S_{ch}(m_2)\right) = 5/7$$

The probability of survival of *schema* is p_s = (1 − probability of being destroyed)

$$p_s(S_{ch}) = 1 - p_d(S_{ch}) \tag{4.7}$$

Generalizing these concepts, the probability of destroying a *schema* is

$$p_d(S_{ch}) = \frac{L(S_{ch})}{(l_s - 1)} \tag{4.8}$$

The probability of survival of a *schema* is

$$p_s(S_{ch}) = 1 - \frac{L(S_{ch})}{(l_s - 1)} \tag{4.9}$$

Let p_c be the *crossover* probability, then

$$p_s(S_{ch}) = 1 - p_c \frac{L(S_{ch})}{l_s - 1} \tag{4.10}$$

The members of the *schema* S_{ch} at the iteration (*iter* + 1) increase at the rate of

$$M(S_{ch}, iter + 1) = M(S_{ch}, iter) . \frac{f(S_{ch})}{f_{av}} \left[1 - p_c \frac{L(S_{ch})}{l_s - 1} \right] \tag{4.11}$$

Mutation is randomly flipping a bit in the chromosome with a probability p_m. If the *schema* is to survive, then all the bits in the fixed positions should be intact. One allele survives mutation with probability $(1 - p_m)$. The alleles in the fixed positions of the *schema* survive with probability $\{(1 - p_m).(1 - p_m) \dots (1 - p_m)\} = (1 - p_m)^{O(\text{Sch})}$. When $p_m \ll 1$, survival probability of the alleles is $\sim (1 - O(S_{ch})p_m)$. Hence the mathematical equation representing the members of the population in iteration (*iter* + 1) based on the members at iteration (*iter*) and the crossover and mutation probabilities is,

$$M(S_{ch}, iter + 1) \geq M(S_{ch}, iter) . \frac{f(S_{ch})}{f_{av}} \left[1 - p_c \frac{L(S_{ch})}{l_s - 1} - O(S_{ch})p_m \right] \tag{4.12}$$

Based on the mathematical discussions given above, the *Schema Theorem* states that short, low-order, above-average *schema* receive exponentially increasing members in succeeding generations.

4.8 Prisoner's Dilemma Problem

GA strikes a good balance between exploration and exploitation. This is proved in the ability of GA to solve the problem of *prisoner's dilemma*. *Prisoner's dilemma* is a game where there are two players A and B. If both Player A and Player B *cooperate* they receive equal payoff each. If both Player A and Player B *defect* they receive a minimal (equal) payoff each. If one of them *cooperates* and the other *defects*, the defector does not receive any payoff whereas the cooperator receives a higher payoff. Two researchers Robert Axelrod and Stephanie Forest explored the application of GA in solving the *prisoner's dilemma* using the *tit-for-tat* strategy. The *tit-for-tat* strategy begins with *cooperation* and later on copies the other player's move, i.e. *cooperation* for *cooperation* and *defection* for *defection*. The strategy involves encoding strings, possibly bit 1 for *cooperation* and bit 0 for *defection*. In each iteration there are four possible outcomes: (i) *cooperation – cooperation*, (ii) *cooperation – defection*, (iii) *defection – cooperation*, (iv) *defection – defection*. A set of three consecutive plays yields $4^3 = 64$ outcomes requiring 64-bit string for encoding each possible outcome. The fitness function is the average payoff of the players A and B. GA was able to discover the *tit-for-tat* strategy to maximize the average payoff. Further to this, a variation was introduced, where players are 'bluffed' into *cooperating* in response to *defection*. GA discovered that players could not be bluffed and reverted to *tit-for-tat* strategy.

4.9 Variants and Hybrids of GA

The **messy genetic algorithm** is a variant of the classical GA in encoding genes with variable length codes that are independent of the position of the gene within the chromosome. In normal GA the gene is identified by its position in the chromosome, whereas in the messy GA the gene is represented by an index to identify its position and an allele. The advantage of this variant is that the genes can occupy any position in the string. Another difference between classical GA and messy GA is the use of a *cut and splice* operator instead of the normal *crossover* operator that facilitates mating between parents of variable length. The **Adaptive Genetic Algorithm** adapts itself during the running of the algorithm. The parameters of the algorithm such as population size, *crossover*, and *mutation rate* could be adaptively modified during the iterations based on the average fitness values or the convergence towards the global optimum.

Self-Organizing Genetic Algorithm organizes itself for encoding and applying operators on the genetic material. The encoding of the chromosomes and application of *selection*, *crossover*, and *mutation* play a crucial role in the performance of the GA. Introducing additional functionality into the GA for these operations makes the GA self-organizing. **Hybrid Genetic Algorithms** combine GA with some other optimization strategy such as the classical methods or swarm intelligence-based algorithms. It has been found that a hybrid combination of GA with some of the other methods gives a better result in terms of the global optimum solution or convergence rate or reduced computational complexity. The hybrid optimization algorithms are available in the literature such as GA-PSO, GA-ACO, and a host of other swarm intelligence algorithms combined with GA. GA has been applied to a wide spectrum of problems over the years since its inception in the 1960s and has been found to give good performance in terms of the optimum solution attained. Some of the notable applications are feature extraction and classification in image processing, image compression, job shop scheduling, and optimization of complex engineering designs.

4.10 Summary

Biological evolution creates individuals, where there is no individual with super fitness but groups of individuals with similar characteristics like a classifier system. The individuals interact with each other. Similarly GA also can be devised to create solutions or evolve as a system, similar to a classifier where the characteristics or features are encoded in strings. In such higher level problems, the strings should be made to represent rules or an hypothesis and offspring will be new rules (evolved from parents) or hypothesis. GA uses a population of search agents (strings) to search for the optimum solution in the search space. This embeds implicit parallelism in the algorithm and helps in searching and exploiting large regions of the search space simultaneously with fewer strings.

GA encodes the parameters of the problem to be solved in the form of strings, whose basic elements could be binary 1s and 0s or some other non-binary alphabet. GA uses an objective or fitness function but does not require the computation of derivatives of the objective function. It uses stochastic rules while searching, by introducing controlled randomness into the algorithm instead of being completely deterministic as in the classical optimization algorithms. The *crossover* operator produces offspring equal in number to the parent population. So *selection* is required to choose the population for the next generation. One method is to choose the population from the offspring only, i.e. parents are replaced with offspring. This approach of choosing only offspring for the next generation is called *generational genetic algorithm*. This increases diversity and prevents convergence to a local minimum but the rate of convergence is reduced. The *elitist* strategy is another technique where the next-generation population is chosen among the total population constituted by the previous generation parents and their children. The fitness value of every member could be used in the selection, selecting the fittest members. One variation in the *elitist* strategy is to limit the number of parents participating in the selection based on fitness values. Randomness or stochasticity in GA occurs in the *selection* of population members for mating and reproduction, choosing the crossover point during *crossover* operation and selecting the chromosomes and the genes within the chromosome for *mutation*. Typical parameters in GA include a population size of 40, number of iterations 20, and mutation rate as 0.005. The encoding and length of the chromosome depends on the problem, and the selection and crossover mechanism is pseudo-random.

References

1. John H. Holland, *Adaptation in Natural and Artificial Systems*, University of Michigan Press, Ann Arbor, MI, 1975 (re-issued by MIT Press, Cambridge, MA, 1992).
2. David E. Goldberg, *Genetic Algorithms in Search, Optimization and Machine Learning*, Addison-Wesley, Reading, Massachussets, 1989.
3. Darrell Whitley, A genetic algorithm tutorial, *Statistics and Computing*, V, pp. 65–85, June 1994.
4. Melanie Mitchell, Genetic algorithms: An overview, *Complexity*, 1 (I), pp. 31–39, September/ October 1995.
5. John H. Holland, Genetic algorithms, *Scientific American*, 267 (1), pp. 66–73, July 1992.
6. Colin Reeves, Chapter 3: Genetic algorithms, In: *Handbook of Metaheuristics, International Series in Operations Research and Management Science*, Michel Gendreau and Jean-Yves Potvin (eds). Springer, Switzerland, pp. 109–139, 2010.
7. Abraham A., Nedjah N., Mourelle L. M. Evolutionary computation: From genetic algorithms to genetic programming, In: Nedjah N., Mourelle L.M., Abraham A. (eds) *Genetic Systems Programming, Studies in Computational Intelligence (SCI)*, Vol. 13, pp. 1–20, Springer-Verlag, Berlin, Heidelberg, 2006.
8. Ulrich Bodenhofer, *Genetic Algorithms: Theory and Applications, Lecture Notes*, 3rd edition, Johannes Kepler University, Linz, October 2003.

5

Genetic Programming

5.1 Introduction

Genetic programming (GP) belongs to the family of evolutionary computational algorithms that can be applied to problems which are difficult to solve with the traditional methods. Friedberg pioneered the work on evolutionary algorithms in 1958 from which genetic programming evolved. The genetic programming technique was proposed by John R. Koza [1] in 1989 and is an extension of the genetic algorithm (GA). Evolutionary algorithms try to mimic the biological process and are based on the Darwinian principle of *survival of the fittest*. Evolutionary algorithms could be applied to problems where heuristic techniques might not produce optimum results. They are suitable for solving practical problems in several domains [2]. Evolution occurs with *survival of the fittest* in populations that compete for existing natural resources. The fit individuals contribute more to the process of reproduction, and hence they are more likely to be members of or produce offspring for the next generation. Evolutionary algorithms are suitable for optimizing unimodal as well as multimodal functions and are simple and easy to implement.

GP is one of the classical evolutionary programming techniques that can be simulated on a computer. The evolutionary algorithms are based on the biological evolutionary processes inculcating the genetic operations of *selection, reproduction, crossover*, and *mutation*. These algorithms are designed to work by evolving towards higher quality solutions that are optimal or near-optimal for a problem. The quality of the solution is related to a numerical value obtained by evaluation of a function that is typically the objective or fitness function related to the problem. The fitness function depends on the problem or application, for the solution of which the evolutionary algorithm is designed. Mimicking evolutionary strategies in nature provides solutions to complex engineering problems in a simple and effective manner. The solution is obtained with reduced time, space, and computational complexity and it is either the global optimum or a good approximation to the global optimum.

GA is the predecessor to GP and has been successfully applied to practical, real-time problems that were difficult for the traditional algorithms to solve. Evolutionary programming (EP) is also a predecessor to GP and was developed in the 1960s by Fogel, Owens, and Walsh. EP mainly applies the *mutation* operator on a finite-state machine (FSM). Here, FSM is a computer program that moves from one state to another in finite time intervals. The change of state takes place based on the present state and the current inputs to the program. EP has a stochastic component inculcated in *selection, reproduction*, and *mutation* and tries to evolve towards better quality solutions to the problem. GP has a population of variable length programs that can be stored using tree, linear, or graph structures in the computer memory. The computer programs evolve and solve problems without being explicitly programmed. LISP is a programming language commonly used in GP because it is simple and supports dynamic data structures. The tree structure is popular for representing programs since it is easy for traversal and evaluation. LISP has gained importance in GP because of its ability to handle tree structures. Moreover, LISP has the advantage of using the same data structure for programs as well as data which makes it easier to manipulate. GP has diverse applications in areas such as financial modeling, predictive modeling, data modeling, data mining, engineering design, feature selection and classification, and so on. The basics of genetic programming, expression trees and their traversals, genetic operators, and the genetic programming algorithm along with the different types of GP have been discussed in the subsequent sections.

5.2 Basics of Genetic Programming

Genetic algorithm is based on a population of individuals that search the space in parallel for the optimum solution. This search process involves the application of genetic operators on the individual members of the population such as *selection, reproduction, crossover,* and *mutation* for producing offspring [3]. *Selection* is selecting the individuals with high fitness values for reproduction (parents), and *reproduction* is the process of producing offspring (children) from parents. *Crossover* is the operator that creates offspring with exchange of genetic material between the two parents. *Mutation* is the operator that introduces diversity into the population by changing the genetic code of an offspring randomly, and thus its character. GA is an iterative algorithm producing members with higher average fitness in succeeding generations than the previous generation. The algorithm stops when the optimum solution based on some predefined criteria is attained or the number of iterations reaches a maximum.

GP is similar to GA in operational procedure. GP generates a population of computer programs from a statement of the problem [4]. The programs are created in LISP (LISt Processor), originally developed by John McCarthy in the 1950s. In LISP, an interpreter can respond directly to programs instead of using a compiler. Another advantage of LISP is that the programs and data can be put in the same data structure which makes manipulation and evaluation easy. An initial population of programs is created using the set of functions and operators along with the set of operands and variables or constants, comprising the parameters of the problem to be solved. The population in GP is a set of computer programs that are created randomly, and the size of the population depends on the problem to be solved. The fitness values of all the members of the population are computed by executing the programs that constitute the population. The *selection, crossover,* and *mutation* operations are applied on the population in every generation to create offspring. This is repeated until the optimum solution is attained or the maximum number of iterations is reached. The member with the best fitness value is the global optimum solution for the problem. The process of *selection* based on fitness values is used for reproduction as well as keeping the population size constant in all generations.

The primitive elements in GP are the functions or operators and the terminals or operands from which the programs are built [5]. Functions process values whereas terminals provide values to functions. Function set comprises logical functions like NOT, AND, OR, XOR, arithmetic functions like ADD, SUBTRACT, MULTIPLY, DIVIDE, scientific functions like EXP, LOG, COS, SIN, statements in a program like ASSIGN, IF-ELSE, WHEN, GOTO, JUMP, REPEAT, or any other function created by the user. The function set chosen for GP should be of sufficient size and include functions that are necessary for solving problems. Too many functions in the set can enlarge the search space, increasing the search complexity. Smaller number of functions might not be sufficient to solve all problems. The terminal set is comprised of variables and constants which are inputs to the functions, or they can be functions without arguments (zero-input functions that return a value without taking an input). Similarly, the number of terminals should be of medium size, neither too large nor too small. On an average, it has been found that approximately 56 functions and 200 terminals can solve most of the real-time problems in finite time. Constants that form terminals in GP can be combined to form other constants, and also functions can be combined to form other higher level functions. Functions must be chosen such that they accept all constants (or variables) as inputs.

The programs in GP are built from the fundamental constructs comprising functions and terminals. The programs are assembled from the defined set of functions and terminals and the rules of the programming language. The programs in GP have to be represented in a suitable manner so that storage and manipulation are easy. Usually programs will not be of the same or fixed length; they vary depending on the application. It is difficult to create programs based on *crossover* and *mutation* that are syntactically correct. The programs can be written in any computer language, but the preferred language is one in which the tree, linear, and graph structures are easy to represent. The tree representation is more popular because storage, traversal, and evaluation are easy. The tree is composed of internal nodes and external nodes that consist of functions (or operators) and variables or constants (or operands) respectively. Internal nodes have branches emanating from them, whereas the terminal nodes are leaf nodes that do not have further branches, i.e. they terminate the branches. GP requires the formulation of an appropriate fitness function, defining rules and elements of the programming language, and interpretation of the programming language. GP incorporates randomness in decision-making in the program. The functions operate on terminals or outputs from other functions or programs and the genetic operators of *selection*, *reproduction*, *crossover*, and *mutation* are used in evolving the programs. The operation of *crossover* can be done on the tree by exchanging subtrees among two trees. The subtrees are chosen randomly just like choosing the crossover site in chromosomes in GA. The next operator is *mutation* where the subtree is chosen and the operator or operand change randomly. This is equivalent to flipping bits in chromosomes in GA. The programs produce a possible solution to the problem when they are run or executed. This is equivalent to evaluation of a fitness function in GA whose value depends on the parameters of the search space. The optimum solution lies in the search space spanned by the population of evolving computer programs. The fitness function is used in *selection* of programs for reproduction and hence in evolution of the population of programs. This evolution is similar to GA except that the programs are executed in order to produce fitness values.

5.3 Data Structures for Genetic Programming

In GP the population of programs is comprised of functions (operators) and terminals (operands). The programs should be assembled in executable form along with rules for execution. The data structures used in building programs in GP reflect on the memory requirements and program execution flow. Commonly used data structures for programs in GP are tree, linear, and graph structures.

The programs can be created as expression trees that produce a candidate solution (fitness value) when executed. The trees are created in place of computer code using a basic set of alphabets that consist of operators and operands. The operators could be simple addition or multiplication, logical functions like OR, AND, or other functions like SIN, LOG, etc. The operands could be variables or constants or another function/program (expression) that represents the parameters of the problem. A typical tree structure that represents an executable program is shown in Figure 5.1.

The program can also be organized as a linear structure with registers holding the values of variables and constants. Corresponding to the tree structure given in Figure 5.1, the set of registers (A, B, C, D, E, F, a, b, c, p, q) can store the values of the variables and constants associated with the program as shown in Figure 5.2a, and the following instructions could be implemented in sequence as shown in Figure 5.2b:

$$A = a + b; B = b - 4; C = \log B; D = p \text{ OR } q; E = C + D; F = A * E$$

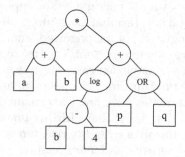

FIGURE 5.1
Typical tree structure for a program.

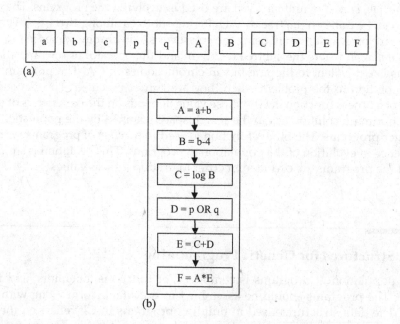

FIGURE 5.2
(a) Registers for holding variables and constants. (b) Sequence of instructions in a linear structure.

A linear structure could be used to hold these instructions that can be executed sequentially. In tree implementation the memory is local (that means a function node can only access its associated terminal or non-terminal nodes; the function + in the left subtree can only access nodes *a* and *b*), whereas in linear structure implementation, the instructions can access any register.

A graph is yet another structure for representing programs in a compact form in GP. Graphs consist of a set of nodes connected by edges. Edges could be undirected or directed. If the edges are directed, they indicate the direction of program flow. The graph uses two types of memory – stack and an indexed memory – to store values used by the program. Each node in the graph operates on data either in the stack or indexed memory and determines which node is the next in the sequence to execute. Every graph has necessarily two nodes, the *start* and *end* node. The program begins execution at the *start* node and terminates at the *end* node. The nodes in the graph are functions that operate on values stored in the stack.

Figure 5.3 shows a simple graph structure in GP. The nodes are executed in sequence beginning from the *start* node and ending with the *end* node. The nodes are executed with data from either stack or indexed memory. The node *a* pushes the value of the variable *a* into the stack, and the node *b* pushes the value of the variable *b* into the stack when they are executed. The node + fetches two items from the stack, performs addition on them, and pushes the result into the stack.

The maximum size of the program should be fixed since it has an effect on memory and execution time. This is physically limited by the number of nodes in the tree or graph or the number of elements in a linear structure. In GP, the population of programs has to be initialized. For tree structures there are two methods of initialization. The sets of functions and terminals are defined. From these two sets, the functions and terminals are chosen randomly and the tree is built. In this method, all subtrees will not be of equal depth since the tree is built in a random manner to some extent. The subtree will stop growing when it is terminated with a branch at the end of which there is a terminal, whereas if a branch contains a function or a non-terminal node, the tree will grow to any depth desired. In an alternative method, the tree is initially built with only function nodes to the required depth and then the terminals are attached to the branches of the tree at the appropriate positions. This makes the tree balanced and of equal depth. This makes the population uniform, and to make it diverse, the technique of initializing the population can be modified. Some members of the population are created with the first method and the remaining members with the second method. The depth of the trees could vary instead of being uniform throughout. This introduces some diversity into the population.

In linear structures, the population is initialized in a similar way. The function set and the terminals along with the registers are chosen initially. Every member of the population is a linear structure that has a header, footer, and a return function. The length of the program is fixed and the functions are chosen randomly to form the structure. The registers from which these functions take the data are also randomly chosen. The entire population is built in this manner. The initial population built in this manner undergoes evolution to form successive generations with evolving (improving) fitness values until the optimum is attained.

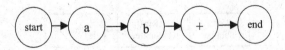

FIGURE 5.3
An example of a graph structure.

The commonly used language in GP is LISP since it is a structured language suitable for such tree representations. The data structure is same for storing functions and variables/data which makes it easy to manipulate and LISP is more suitable for tree representation of programs. But it is difficult to check syntax and process tree structures that include numerals, strings, logic functions, etc. To overcome this limitation of LISP and allow flexibility, a general approach has been proposed by A. Geyer-Schulz to represent programs by syntactical derivation trees. This representation is with respect to Backus–Naur Form (BNF) that has syntactic rules and works for any context-free language. The tree is recursively derived from the grammar using the syntactic rules. BNF can be given as input the rules and syntax of any programming language. The language should be simple and powerful enough to solve the problems, reducing the search space as well as the complexity.

5.4 Binary Tree Traversals

Tree traversal is visiting each node in a tree once. The order in which these nodes are visited leads to three types of traversal known as *PreOrder*, *InOrder*, and *PostOrder* traversals. In *PreOrder* traversal, the order of visiting nodes is root → left subtree → right subtree as shown in Figure 5.4. In *InOrder* traversal, the order of visiting nodes is left subtree → root → right subtree as shown in Figure 5.5. In *PostOrder* traversal, the order of visiting nodes is left subtree → right subtree → root, as shown in Figure 5.6. This is done recursively for all the subtrees until all the nodes of the tree have been visited once.

When a tree is traversed using one of the above three techniques, the same traversal is repeatedly done for all the subtrees until the entire tree has been traversed. When the tree is a binary expression tree, the three traversals yield expressions in three different forms. The three types of traversals yielding *prefix*, *infix*, and *postfix* expressions are shown in Figure 5.7a, b and c respectively. *PreOrder* traversal yields *prefix* expression, *InOrder* traversal yields *infix* expression, and *PostOrder* traversal yields *postfix* expression. In *prefix*

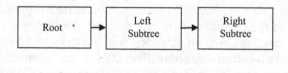

FIGURE 5.4
PreOrder traversal.

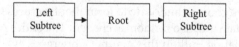

FIGURE 5.5
InOrder traversal.

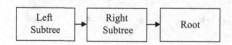

FIGURE 5.6
PostOrder traversal.

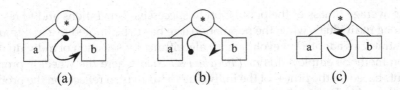

FIGURE 5.7
(a) *ab (b) a*b (c) ab*.

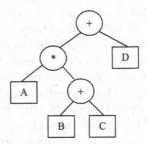

FIGURE 5.8
Binary expression tree.

notation, the operator appears before the operands. In *infix* notation, the operator appears between the operands. In *postfix* notation, the operator appears after the operands. As a simple example, consider the expression a*b, where the multiplication operator appears between the operands. This is the standard *infix* notation normally used in all the evaluations. In *prefix* notation, the expression becomes *ab, and in *postfix* notation, the expression becomes ab*. When the expressions are evaluated with values assigned to the variables, all three types produce the same result.

In a binary expression tree, the internal nodes are the functions or operators like +, −, *, /, and the terminal nodes are the operands or variables upon which the functions operate. The evaluation of any expression tree gives a value that is the same as that obtained by evaluating the mathematical expression given in one of the three forms. These notations for writing mathematical expressions to be evaluated can be obtained from the binary expression tree by appropriate parsing.

As an example consider the binary expression tree in Figure 5.8.

The tree is traversed in *PreOrder, InOrder,* and *PostOrder,* yielding the expressions in *prefix, infix,* and *postfix* notation respectively. The expressions obtained by parsing the binary tree in Figure 5.8 are given below:

PreOrder traversal produces the *prefix* expression: + * A + B C D

InOrder traversal produces the *infix* expression: A * (B + C) + D

PostOrder traversal produces the *postfix* expression: A B C + * D +

5.5 Genetic Programming Operators

Selection is choosing the individual members of the population, either for reproduction or inclusion into the next generation based on their fitness values. *Selection* is done mainly to improve the average fitness of the population in successive generations and to resolve competition among individuals when the population size has to be limited to a maximum. There are several strategies adopted in evolutionary algorithms for selection of individuals. One of the common methods employed is *roulette wheel selection* where the wheel is proportioned into segments based on the fitness of the individual. This in turn reflects on the probability of selection of the individual. Individuals with larger segments have higher probability of being selected and individuals with smaller segments have lower probability of being selected. The second method is *tournament selection* where a selection of individuals (*tournament* size is number of individuals chosen) is made randomly. Competitions are conducted among those chosen, and the winning individuals with higher fitness become parents on which *crossover* or *mutation* could be applied. In *ranking selection*, the individuals are ranked according to their fitness and those with higher fitness get selected. The probability of selection is proportional to their rank in the list. In *truncation selection*, let there be p number of parents breeding to produce o number of offspring making the total population ($p + o$). Out of this total population, p number of best individuals are selected to become parents for the subsequent generation.

Crossover is implemented in GP by selecting subtrees from the two parents and interchanging them. The selection of parents can be done using one of the methods outlined above such as *roulette wheel selection, tournament selection,* etc. Figure 5.9a and b show two parent trees and the selected subtrees identified for *crossover* operation. Figure 5.10a and b show the two offspring produced as a result of the *crossover*. The subtree selection is done randomly like choosing the crossover point in a chromosome.

Two trees (parents) are chosen either randomly or based on their fitness values. Their subtrees are again chosen randomly and swapped to produce two children or offspring. In *linear* structures, the segments of the two parents are chosen randomly and exchanged. This creates two children. In *graph-based* structures, the selected parents are divided into two sets of nodes and edges, called *fragments*. The edges which connect nodes within a fragment are internal whereas those that connect to nodes in the other fragment at one end are

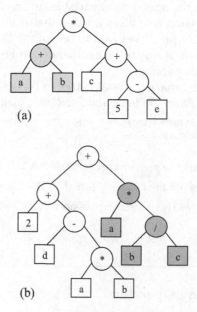

FIGURE 5.9
Parents with subtrees chosen for *crossover*.

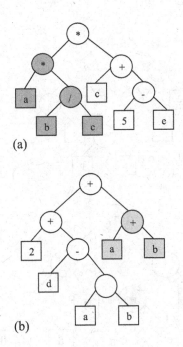

FIGURE 5.10
Offspring resulting from *crossover*.

external. The nodes are labeled as output if they are a source of data to any outgoing edge and labeled as input if they are the destination of data incoming from an edge. The fragments are swapped and the nodes are interconnected through external edges accordingly.

Mutation is the operation where one subtree can be replaced with another one or one function (operator) or terminal (operand) can be replaced with another one. In *mutation*, the operators or functions in inner (internal) nodes are chosen randomly in a tree and replaced with other functions. Similarly, the terminal (external) nodes are replaced with some other variables or constants. Extending this concept further, an entire subtree can be replaced with another subtree.

Figure 5.11 shows how *mutation* takes place by replacing randomly selected existing operators and operands with some other operators and operands. Figure 5.11a shows the expression tree selected for *mutation* with the identified operators and operands. Figure 5.11b shows the expression tree after the *mutation* operation has taken place. Figure 5.12 shows an example of *mutation* where one subtree is replaced with another one. Figure 5.12a shows the expression tree and the identified subtree, and the Figure 5.12b shows the expression tree after *mutation*.

The probability of *mutation* in a population is very low, typically less than 0.1. In *linear GP*, one member of the linear structure is chosen randomly. The chosen operator or constant is changed, again in a random manner, to create a new member by *mutation*.

When *crossover* is done on trees by exchanging subtrees, the resulting tree might not be syntactically correct. In order to overcome this problem, the exchange of subtrees should be done between those that start from the same non-terminal symbol, so that the resulting trees do not violate syntactical rules. For mutation, a randomly chosen subtree is replaced with another one, and in a similar manner, the non-terminal symbol which is the root of the subtree is chosen for replacement.

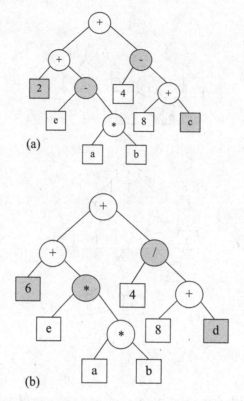

FIGURE 5.11
Mutation on operands/operators (a) before *mutation*, (b) after *mutation*.

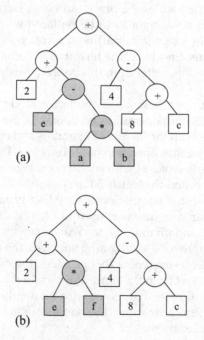

FIGURE 5.12
Mutation on subtree (a) before *mutation*, (b) after *mutation*.

Fitness function is a mathematical function (or expression) that is problem-specific and is used in evaluation and selection of individuals in a population. This fitness value is used in the selection of individuals for reproduction so that individuals with higher fitness will contribute more to the next-generation population. Also, individuals with lower values of fitness could be removed from the population. The fitness values are computed with a training set of data applied as input to the programs and obtaining the output. The outputs of the programs that are executed from the inputs provided are the fitness values. The fitness function should be chosen appropriately (either maximization or minimization) so that it is an indication of the result (correctness of the output) of the program. Fitness functions that are squared or scaled can lead to amplified or damped values which can lead to better results.

5.6 Genetic Programming Algorithm

The algorithm for GP starts with the problem definition and constraints, if any. The objective or fitness function has to be defined based on the optimization problem. It could be either maximization or minimization of the objective, as the case may be. The corresponding mathematical equations for the fitness function and constraints have to be formulated. Since GP uses a population of programs, the parameters such as number of members in the population N, maximum program size, set of functions, and terminals have to be chosen carefully. In addition, the genetic operators (*crossover* and *mutation* probabilities), maximum number of iterations, and termination criteria, if any, have to be defined. An initial population of programs has to be created with the set of defined functions (operators) and terminals (operands). The fitness value of the entire population has to be computed by executing the programs.

Parents are selected using a suitable strategy, and offspring are created by applying the defined genetic operators. The fitness values of the entire population are calculated, and the next-generation population is chosen based on higher fitness values and the weaker ones are discarded. This keeps the population size constant at N in every generation. The processes of *selection*, *crossover*, and *mutation* (optional) are repeated until the maximum number of iterations is reached or the optimum solution is found or termination criterion is attained. To implement the tree structure for representing and editing the programs, GP requires appropriate data structures that are easily stored and manipulated. The performance of GP depends on the set of functions (operators) and terminals (operands) chosen, the initial population and its size, depth of the trees, and the selection of parents in producing offspring.

There are broadly two ways of designing a GP algorithm. One is *generational* GP and the other is *steady-state* GP. In *generational* GP, each generation of the population is distinct. An initial population of individuals is created, and they are evaluated for their fitness values. The parents are selected using one of the selection criteria, and the genetic operators like *crossover* and *mutation* are applied. The resultant offspring form the new population, with children replacing parents. This is the next generation of individuals that entirely replaces the previous generation. If the termination criterion is met the GP run stops; otherwise it continues until the maximum number of generations is reached. The individual with the best fitness value is the optimum solution to the problem. In *steady-state* GP, there are no distinct generations of population as in *generational* GP. The process of GP is a continuous flow of operations. The initial population of individuals is created. Members of the population are selected randomly to participate in the *tournament* and their fitness is evaluated. The winners are chosen using a *selection* strategy, and the genetic operators are applied on them. The new members created replace the losers of the tournament. This is repeated until the termination criterion is met or the maximum number of iterations (generations) is reached. The individual with the best fitness value is the optimum solution to the problem.

The initial population is usually generated with a uniform distribution, and in GA it is easy to generate an initial population of binary strings. In GP, an initial population of programs has to be generated with the maximum size of the program being one of the important parameters. Assuming tree structure for the programs, some rules have to be adopted since it is not as easy as generating a random population of binary strings with a uniform distribution. The maximum number of nodes in the tree or the maximum depth of the tree has to be fixed for the problem. The trees can be generated recursively with the defined set of terminal and non-terminal symbols using the rules of grammar. In the *grow* method, the functions and terminals are selected randomly from the defined set and the tree grows. If a terminal node is encountered at the end of an edge or branch, the branch terminates. The trees generated by this method are not uniform or regular. In the *full* method, the function nodes are selected to form the tree of the required depth and then the terminals are chosen and attached at the appropriate branches. This method creates a population of trees that are uniform. To improve the diversity of the population, the maximum depth of the trees is fixed and half the tree population is generated using *grow* method and the other half using *full* method.

5.7 Pseudocode

Initialization

    Population size N

    Define objective function $f(X)$

    Define set of functions and terminals

    Initialize the population of programs

    Compute the fitness of the initial population

    Choose the genetic operators

    Termination criteria, if any

    Maximum number of iterations *MaxIter*

    iter = 1

while (*iter* ≤ *MaxIter*) **do**

    Select parents for mating and reproduction

    Apply *crossover* and *mutation* (optional) to produce offspring

    Strategically choose the next generation population (N)

    Compute the fitness values of the entire population

    if termination criteria met **exit, else** continue

    iter = *iter* + 1

end while

Program with highest fitness value is the global optimum solution

Flowchart

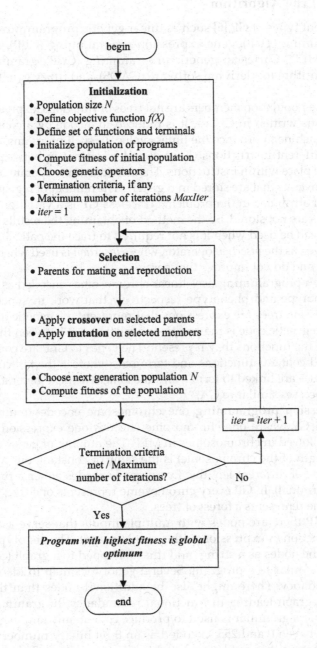

5.8 Variants of the Algorithm

There are different types of GP [6] such as linear genetic programming (LGP), traceless genetic programming (TGP), gene expression programming (GEP), multi-expression programming (MEP), Cartesian genetic programming (CGP), grammatical evolution (GE), genetic algorithm for deriving software (GADS), and fuzzy genetic programming (FGP).

In **linear GP**, the population members are not trees or functions (expressions to be evaluated) but programs written in C/C++. It is reduced to a code by a compiler that can be evaluated on a machine to produce the fitness value. In such programs, *crossover* can take place between different instructions (lines of code) and not in between an instruction. *Mutation* can take place within instructions. These instructions operate on predefined sets of variables or constants that are stored in registers. The number of registers used depends on the number of attributes in the problem. **Traceless GP** is a variation of GP where the evolved programs are not stored. So the method of obtaining the results cannot be traced, and this method can be used when it is not required to trace the path. TGP uses the *crossover* operator as well as the *insertion* operator, where *insertion* is used when the lines of code are very complex and do not improve the search.

Gene expression programming uses linear chromosomes encoded as expression trees. They are the genotype and phenotype respectively that work in synchronization with each other. Expression trees are expressions that represent the genetic information of the chromosome. The genetic code is the one-to-one relationship between the elements of the chromosome and the functions they represent. The genes in GEP are composed of a head and tail. The head contains functions and terminals whereas the tail contains terminals only. The GEP genes are linked to form a chromosome. The next-generation population is created using processes similar to GA.

In **multi-expression programming** one chromosome encodes multiple expressions, unlike in normal GP where one chromosome encodes one expression. Variable length expressions are allowed in chromosomes in MEP. The number of genes making up every chromosome (length of the chromosome) is the same or constant, but within a gene, the expressions can be of different lengths. Every gene encodes either a function (operator) or an operand (terminal). In GP every chromosome represents one tree, whereas in MEP, every chromosome represents a forest of trees.

In **Cartesian GP** there are nodes with multiple inputs that serve as parameters in a mathematical function or expression with a single output. The genotype represents the functionality of the nodes as a string, and this is mapped to a graph (connected nodes) that can execute similar to a program. Several genotypes map to identical genotypes, leading to redundancy. There might also be additional nodes than the ones that are connected to the graph leading to functional redundancy. In **grammatical evolution** the *Backus–Naur form* grammar is used to produce code in any language. The genotype is any number between 0 and 255, encoded as an 8-bit binary number. The phenotype is a computer program that can be executed to produce output. The program is generated by genotype–phenotype mapping that is deterministic. The operators in GA are applied here on the genotypes. In **genetic algorithm for deriving software** the genotype and phenotype are different from each other. Genotypes are integers, and they are used to generate phenotypes that are programs. **Fuzzy genetic programming** is a synergy of BNF and GP. The fuzzy rules of the language are specified in BNF and given as input to GP.

5.9 Summary

Genetic programming is an integration of biological evolution and computer science with inclusion of artificial intelligence and machine learning. GP involves computer programs that evolve which requires computation-intensive resources, and it is an advanced concept compared to other evolutionary algorithms. The data structures for representing programs in GP become important with respect to the space and time complexity of the algorithm. The possibility of using different data structures such as trees, graphs, and other linear structures gives an added advantage in GP. The memory required to store the population of programs and the data upon which the program operates are dependent on the data structure used for storing the programs. The fitness function of every member of the population has to be obtained by executing the programs which could use computational resources depending on the length and structure of the individual program. This could also be an advantage in GP since it gives the user some flexibility in creating variations among the population members instead of their being identical.

Population size is one of the important parameters in GP in searching for the optimum solution. A smaller population size evolves faster, but a large population is required for solving complex problems and improving the diversity of the search. Typical population size could be less than 1000 for small problems, but it can increase depending on the size of the problem. The maximum number of generations can initially start from 20 and could go up to 100 or even beyond that if the problem requires a higher number of iterations. The set of terminals and functions should be small so that the computations are reduced, but it should also be able to accommodate non-linear problems. The set of functions should be chosen meticulously to be specific to the problem. The right balance of crossover and mutation operations is essential to ensure the diversity of the population and also faster convergence. Typically the genetic operations could have a tournament size of 4 (if tournament selection is applied), 90% crossover and 10% mutation. For tree representations, the choice of the depth of the tree and number of nodes (external and internal) is also important in the computations required. The landscape of the objective or fitness function is dependent on the population of programs that evolve, making the landscape dynamic. The inherent parallelism in GP and its rate of convergence play an important role in the expanding applications of GP. It can accommodate problems with smaller population size as well as data-intensive applications. In the future, GP can be hybridized with other evolutionary or swarm intelligence algorithms to improve performance. The effect of parameter tuning on the performance of GP is to be explored.

References

1. John R. Koza, Hierarchical genetic algorithms operating on populations of computer programs, *Proceedings of the 11th International Joint Conference on Artificial Intelligence (IJCAI '89)*, Vol. 1, Detroit, MI, August 1989.

2. John R. Koza, *Genetic Programming: On the Programming of Computers by Means of Natural Selection*, MIT Press, Cambridge, Massachusetts, 1992.

3. Ulrich Bodenhofer, *Genetic Algorithms: Theory and Applications, Lecture Notes*, 3rd edition, Johannes Kepler University, Linz, October 2003.

4. Koza J.R., Poli R., Genetic programming. In: Burke E.K., Kendall G. (eds) *Search Methodologies*, Springer, Boston, MA, pp. 127–164, 2005.
5. Wolfgang Banzhaf, Peter Nordin, Robert E. Keller, Frank D. Francone, *Genetic Programming, An Introduction*, 1st Edition, Morgan Kaufmann Publishers Inc., New York, 1998.
6. Abraham A, Nedjah N, Mourelle L M, Evolutionary computation: From genetic algorithms to genetic programming, In: Nedjah N., Mourelle L.M., Abraham A. (eds) *Genetic Systems Programming, Studies in Computational Intelligence (SCI)*, Vol. 13, pp. 1–20, Springer-Verlag, Berlin, Heidelberg, 2006.

6

Particle Swarm Optimization

6.1 Introduction

Particle swarm optimization (PSO) was developed by James Kennedy and Russell C. Eberhart in 1995 at Purdue University. PSO has its roots in the flocking behavior of swarms of birds combined with the principles of evolution [1]. A flock of birds fly together without bumping into each other, keep optimum distance with their neighbors, and execute their activities collectively. Figure 6.1 shows a flock of birds exhibiting collective behavior at the Kadalundi Bird Sanctuary.

Several scientists have tried to simulate the flocking of birds based on their behavior of keeping (flying) together, scattering, changing direction without bumping into each other, and all these being carried out synchronously. PSO is the first population-based swarm intelligence algorithm that is modeled on the flocking behavior of birds. Since there are multiple agents or particles searching in the space in parallel, the algorithm has inherent parallelism that makes it efficient.

PSO is an optimization algorithm proposed for linear as well as non-linear optimization problems that are either constrained or unconstrained. It does not require the computation of derivatives and is suitable for continuous as well as discrete combinatorial optimization problems. PSO can be applied to continuous, discrete, and mixed search spaces containing single and multiple optima, thus making it suitable for unimodal and multimodal functions. PSO is easy to implement [2] and the function need not be continuous and derivatives are not required to be computed. PSO is a population-based technique where swarms of particles are moving in the search space. An objective or fitness function has to be defined that is appropriate for the problem whose value is obtained by evaluating at different positions in the search space. Each particle in the search space represents a potential solution to the problem. It is an iterative algorithm where the maximum number of iterations has to be specified or a stopping criterion has to be defined, as appropriate. The algorithm requires less memory, is iterative and quite fast.

PSO differs from other evolutionary algorithms such as genetic algorithm (GA) in the manner of evolution of the population. In GA the population changes with every generation, with a mix of the older members along with new offspring created, but the population size remains constant. In PSO the population of particles remains the same in every iteration throughout the run of the algorithm. The same population carries from the first to the last iteration, hence there is no concept of *survival of the fittest*. PSO is an algorithm that is related to GA in using a population of particles and also based on evolutionary principles since the solutions evolve iteratively [3]. PSO is simple to implement and effective for different types of objective functions over a wide range. The swarm exhibits group dynamism and flocking behavior. It also imitates human social behavior, wherein individuals

FIGURE 6.1
Flocking birds at the Kadalundi Bird Sanctuary. (Author: Dhruvaraj S, originally posted to **Flickr,** CC BY-SA 2.0. https://creativecommons.org/licenses/by/2.0/deed.en.)

interact with each other and update themselves based on social interactions [4]. Humans learn from their own as well as the experience of others, and the collective intelligence of swarms is similar to this. PSO combines local as well as global search which can be modeled as intensification and diversification, where intensification is for exploitation and diversification is for exploration.

The uniqueness of PSO lies in modeling the algorithm as flying of particles through the search space [5], which is a d-dimensional hyperspace. The particles acquire velocities, accelerate towards better positions, and finally reach the globally best position. PSO has several candidate solutions in the search space that are represented by particles or birds flying through the search space. This makes it possible to search known as well as unknown regions in the space. In each iteration, the fitness values of each of the candidate solutions are computed based on their positions in the search space. The solutions are evaluated with the objective function whose inputs are the position vectors of the particles. The initial positions of the particles (based on which fitness is calculated) are randomly chosen, and the population size is chosen based on the problem.

The position of the bird or particle in the search space determines its fitness value, and the present position and velocity together determine its next position and new velocity. The velocity should be a tradeoff between being too high and too low. If it is high, the particle can go past the optimum solution, and if it is low, it might converge on a local optimum. Each particle has a best position that it has achieved so far (best fitness value) called *personal best* (P). To store this information requires memory for every particle. The best position among all the particles in the swarm is called *global best* (G), and every particle tries to move towards the *global best*. This will be updated with every iteration. The algorithm is iterative in nature, and the search process is repeated until either the stopping criterion is attained or the maximum number of iterations is reached.

6.2 Swarm Behavior

Swarm behavior refers to the collective behavior exhibited by animals, birds, or insects when they involve themselves in some activity [6]. The entities of the swarm move together, forage, or migrate towards a particular direction in a disciplined manner. This swarm behavior has been utilized in the development of several of the nature-inspired optimization algorithms of which the PSO is the pioneering development based on flocking of birds. A flock of *lesser flamingos* flying together is shown in Figure 6.2.

The birds behave collectively in a self-organized manner and the flock is decentralized. They fly in unison but there is a random component, which is more apt in modeling their behavior as a flock. This randomness makes it realistic. The members of the population interact with each other and with the environment. This leads to a global behavior pattern which is not written down as a rule. The underlying group dynamics of flocking birds is assumed to be based on three rules: (i) Face the same direction as the other birds, (ii) keep near the other birds, (iii) do not bump into any other bird in the flock. These rules have been framed by Reynolds in his 1987 paper [7] as simple rules of the flocking model:

1. Collision avoidance – members of the flock do not collide with each other.
2. Velocity matching – all birds fly at the same speed.
3. Flock centering – try to move towards the center of the flock.

The members of the flock benefit from the experience of others and their own past experience in the search for food. There are advantages of collective foraging as well as competition in the search for food. The advantages outweigh the disadvantages when the food is available in a scattered manner. The social sharing of information among cospeciates is an evolutionary advantage, and this is the underlying fundamental principle of particle

FIGURE 6.2
Flock of *lesser flamingos* flying in formation. (Author: Nikunj Vasoya – own work, CC BY-SA 3.0. https://creativ ecommons.org/licenses/by-sa/3.0/deed.en.)

swarm optimization. Human social behavior is similar but not same as that of birds and animals. Birds and animals keep together to avoid predators, forage for food, and look for mates. Humans have a cognitive component, and they do not act in unison but their attitudes and beliefs are in accordance (conformance) with their peers. *Change* in human social behavior is equivalent to *movement* in bird behavior.

The elements of the swarm have their own personal manifestation as well as the manifestation of the swarm. These are referred to as cognitive and social behavior respectively. The members of the swarm behave based on their own past as well as that of the entire swarm. The position and velocity change based on their own past behavior as well as collective social behavior of the swarm. They move around in the search space which is also the solution space. Each particle/bird has a position and velocity that vary with time. Birds avoid predators, look for food, and keep near their neighbors but do not collide with each other or any other entity. The birds move in the same direction while looking for food or mates or during migration. A flock of *barnacle geese* flying together in formation during autumn migration is shown in Figure 6.3. When birds fly together as a flock, their velocities have to be adjusted to be the same or close to those of their neighbors. The movement has to be synchronous and in the same direction with different positions within the same flock. This can be modeled with some randomness introduced into the movement of the birds. The swarm has to have memory so that it remembers its previous best position.

The five characteristics or principles on which swarm behavior is modeled are [8]:

(i) *Proximity* – population should be able to carry out simple space and time computations.

(ii) *Quality* – population should be able to respond to quality factors in the environment.

(iii) *Diversity* – population should not commit its activities along excessively narrow channels.

FIGURE 6.3
Flock of *barnacle geese*. (Author: Thermos – own work, CC BY-SA 2.5 https://creativecommons.org/licenses/by-sa/2.5/deed.en.)

(iv) *Stability* – population should not change its mode of behavior every time the environment changes.

(v) *Adaptability* – population should be able to change its behavior mode when it is worth the computational price.

The underlying paradigm of PSO is computations carried out in a d-dimensional search space over a sequence of time intervals. This is the first principle of swarm intelligence. The members of the population respond to the quality factors *personal best* and *global best*; this is in conformance with the second principle. The diversity specified in the third characteristic occurs because of allocation of response between *personal best* and *global best*. The population changes state only when *global best* changes; this makes it stable, and the fact that the population does change when *global best* changes makes it adaptive. This is in conformance with the fourth and fifth principles stated above.

6.3 Particle Swarm Optimization

The algorithm begins with a problem statement and associated constraints, if any. The population size N of the swarm and the objective or fitness function are defined based on the criteria to be optimized in the problem. The population of particles or agents is distributed uniformly throughout the search space with their positions and velocities chosen randomly within the defined boundaries. The search space is assumed as d-dimensional, so each particle position and velocity is represented by a d-dimensional vector. The velocity of the particles may be initialized to zero or to some other value within defined bounds. The fitness values for the swarm of particles at their current positions are calculated, and this is their *personal best* since this is the initial fitness value of the swarm. The maximum fitness value among the entire swarm is the *global best*. The particles are accelerated towards the *global best* position of the swarm since their own positions are the current *personal best*. The new fitness values of the particles are calculated based on their new positions. The *personal best* and *global best* of the swarm are updated among all the positions attained by the particles so far. During exploration of the search space in this manner, the particles find solutions which may correspond to either local or global optima. This process is repeated iteratively until the global optimum solution is attained or a maximum number of iterations is reached or a stopping criterion is met [9].

6.3.1 Algorithm

Let $X_i^{iter} = [x_{i1}^{iter}, x_{i2}^{iter}, ..., x_{id}^{iter}]$ be the *ith* particle in the d-dimensional search space in iteration indexed by the variable *iter* and N be the population (swarm) size, where $i = 1, 2, ..., N$. Let $P_i^{iter} = [p_{i1}^{iter}\ p_{i2}^{iter} ... p_{id}^{iter}]$ be the *personal best* position of the *ith* particle, $G^{iter} = [g_1^{iter}\ g_2^{iter} ... g_d^{iter}]$ be the *global best* position of the swarm, and $f(X_i^{iter})$ be the objective or fitness function evaluated for the *ith* particle in iteration *iter*. Each particle has a fitness value based on its position in the search space as obtained by evaluation of the fitness function. The highest fitness value attained by the particle so far in some position of the search space is the *personal best* position of the particle. The highest fitness value attained so far in some position of the search space among all the particles of the swarm is the *global best* position of the swarm. The velocity and position update equations for the particles are given by Equation 6.1 and Equation 6.2 respectively.

$$V_i^{iter+1} = w_v V_i^{iter} + c_1 R_1 \left(P_i^{iter} - X_i^{iter} \right) + c_2 R_2 \left(G^{iter} - X_i^{iter} \right) \tag{6.1}$$

$$X_i^{iter+1} = X_i^{iter} + V_i^{iter+1} \tag{6.2}$$

where $V_i^{iter} = [v_{i1}^{iter}\ v_{i2}^{iter} \dots\ v_{id}^{iter}]$, $R_1 = [r_{11}\ r_{12} \dots\ r_{1d}]$, $R_2 = [r_{21}\ r_{22} \dots\ r_{2d}]$.

These equations are for the *ith* particle in general. Since the search space is *d*-dimensional, the position and velocity of every particle are *d*-dimensional vectors, and they have to be updated for every dimension. The velocity and position update equations for the *jth* dimension of the *ith* particle are given by Equation 6.3 and Equation 6.4 respectively.

$$v_{i,j}^{iter+1} = w_v v_{i,j}^{iter} + c_1 r_{1j} \left(p_{i,j}^{iter} - x_{i,j}^{iter} \right) + c_2 r_{2,j} \left(g_j^{iter} - x_{i,j}^{iter} \right) \tag{6.3}$$

$$x_{i,j}^{iter+1} = x_{i,j}^{iter} + v_{i,j}^{iter+1} \tag{6.4}$$

$i = 1, 2, \dots, N; j = 1, 2, \dots, d$

The vectors in the above equations are *d*-dimensional since the search space is assumed as *d*-dimensional hyperspace. The maximum number of iterations for the algorithm is represented as *MaxIter* and a stopping criterion for the algorithm could also be defined, such as a threshold ε. The variable w_v is the inertia coefficient, and a smaller value for w_v accelerates the particle movement whereas a larger value for w_v dampens the movement. The inertia coefficient is responsible for the movement of the particle in any direction in the search space. If the particles move faster it will result in faster convergence of the algorithm, and if the particles move slowly it will result in slower convergence and more exploration of the search space.

The second term in the velocity update equation $c_1 R_1 (P_i^{iter} - X_i^{iter})$ is the cognitive component that causes the *ith* particle to move towards the best positions found by itself so far. This inherently leads to memory being included in the particle so that it can return to its better positions found in the past. The constants c_1 and c_2 influence the maximum step size the particle can take in the direction of the *personal best* and *global best* in any iteration, so they are called acceleration constants or coefficients. The coefficient c_1 is the cognitive coefficient, and its value determines the step size of the particle taken towards its *personal best* position. The third term in the velocity update equation $c_2 R_2 (G^{iter} - X_i^{iter})$ is the social component and determines the movement of the *ith* particle towards the best positions found by the swarm so far. The coefficient c_2 is the social coefficient, and its value determines the step size that the particle takes towards the *global best* position found by the swarm so far. The coefficients R_1 and R_2 are random numbers that introduce a stochastic component in the movement of the particles of the swarm. This makes it appear that the particles move in a pseudo-random manner towards the *personal best* and *global best* positions. If only the cognitive term is included, the performance will be poorer since there is no interaction between particles. If only the social term is included, the performance will either be superior or inferior to the performance with both cognitive and social terms included, depending on the problem to which it is applied. The parameters w_v, c_1, and c_2 could be in the range $0.8 \leq w_v \leq 1.2$, $0 \leq c_1 \leq 2$, $0 \leq c_2 \leq 2$ which has been found to be satisfactory for most applications. The actual values are chosen based on the problem to be solved. The values of r_{1j} and r_{2j} ($j = 1, 2, ..., d$) are randomly chosen in the range ($0 \leq r_{ij} \leq 1$ and $0 \leq r_{2j} \leq 1$), and they are regenerated every time the velocity is updated. This brings in the stochastic component into the algorithm where the random component is introduced into the trajectory of the particle as it flies towards its *personal best* and *global best* positions.

Let the search space be bounded by $[-X_{max}$ to $+X_{max}]$. The particles must move in the search space within this range and not go beyond this. So a technique called velocity clamping is proposed to limit the maximum velocity of each particle. The bounding limits for the particle velocity are $[-V_{max}$ to $+V_{max}]$, where $V_{max} = k \cdot X_{max}$. k is the velocity-clamping factor taking values in the range $0.1 \leq k \leq 1.0$. In many of the optimization tasks the search space is bounded by $[X_{min}$ to $X_{max}]$ instead of $[-X_{max}$ to $+X_{max}]$. For such problems, the maximum velocity is given by $V_{max} = k \cdot (X_{max} - X_{min})/2$. As mentioned earlier, the velocity of the particles is clamped to the limits $[-V_{max}$ to $+V_{max}]$ where $V_{max} = k \cdot X_{max}$. In the velocity update equation if the magnitude of the new velocity V_i^{iter+1} is less than V_{max} then this value is the new velocity; otherwise if it exceeds this limit, it is clamped to $\pm V_{max}$. If the velocity is not bounded within limits, the particles will fly out of the search space.

The velocity and position of the particles are updated according to Equation 6.1 and Equation 6.2 in every iteration until the algorithm converges. This process is repeated until the stopping criterion is met or the maximum number of iterations is reached. At the end of the iterations, the current *global best* position (fitness function evaluated at the *global best* position) is the global optimum solution to the problem. Some of the stopping criteria used are maximum number of iterations reached, target fitness value of the objective function is attained, no improvement is observed over a number of iterations, normalized swarm radius is close to zero, etc. The optimum solution is attained when the *global best* position is the global optimum. When the global optimum is not reached within a preset number of iterations or the swarm diverges, the algorithm is deemed to have failed. The diverging of the swarm is controlled by the parameter V_{max}. The inertia parameter w_v has to be selected carefully and may be decreased as the iterations progress. This makes the algorithm go from exploration (diversification) to exploitation (intensification) mode. If the inertia weight is large, the search is global, and if the inertia weight is small, the search is local. The particle velocities have to be clamped to a maximum of V_{max}. If the velocity is too large, the particles will fly past the optimum solution; if the velocity is too small, the particles might get stuck in local optima. The PSO algorithm is stable and adaptive. It is stable because it changes state only when the *personal* or *global best* positions change. It is adaptable because it changes state when the *global best* changes. It has a diverse response between *personal best* and *global best*.

6.3.2 Pseudocode

Initialization

Population (swarm) size N

Define objective function $f(X)$ of dimension d

Initial positions and velocities of the particles

Compute fitness values of the particles

Initial *personal best* and *global best*

Parameters: inertia weight w_v, coefficients c_1 and c_2

Random parameters R_1 and R_2

Maximum number of iterations *MaxIter*

Stopping criteria, if any
iter = 1
for *iter* = 1 **to** *MaxIter* **do**
    Update velocity and position of the particles
    Calculate fitness values for all the particles
    Update *personal best* and *global best*
    if stopping criteria met **then** exit, **otherwise** continue
end for
global best **is the optimum solution**

Flowchart

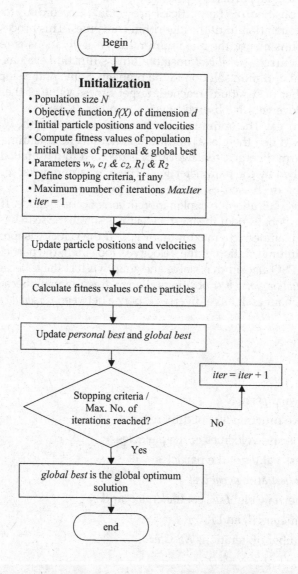

6.4 Variants of the Algorithm

In the original PSO algorithm proposed by James Kennedy and Russell C. Eberhart in 1995, the inertia coefficient w_v is assumed as one, since it is not present in the velocity update equation (Equation 6.1). It has been included in the modified particle swarm optimizer [10] proposed by Shi and Eberhart in their 1998 paper. The inertia component w_v balances the local and global search and mostly it is a constant but it could also be a function. Based on experimental results it is found that when w_v is small (<0.8) PSO is like a local search algorithm and finds the global optimum fast if it is within the initial search space. When w_v is large (>1.2) PSO is like a global search algorithm and explores new areas and hence convergence takes time. When w_v is medium ($0.8 \leq w_v \leq 1.2$) PSO finds the global optimum in a moderate number of iterations. The larger the value for w_v the lesser is the dependency of the algorithm on the initial population. Gradually reducing the inertia weight in a linear manner with each iteration gives the PSO exploration as well as exploitation capabilities. If the first term in the velocity equation is not present, the particles will fly towards the *personal* and *global best* of the swarm and the search space will tend to shrink. If the global optimum is within the initial search space, it has a chance of being found, otherwise not. Therefore, the solution depends on the initial population. With the inclusion of the first term in the velocity update equation, the search space expands. The inertia coefficient balances the local and global search and hence the exploitation and exploration abilities of the algorithm [11].

In one of the modifications to the PSO algorithm proposed by Clerc [1999] a constriction factor has been introduced into the velocity update equation [12]. The modified velocity update equation is given in Equation 6.5.

$$V_i^{iter+1} = K\left[V_i^{iter} + c_1 R_1\left(P_i^{iter} - X_i^{iter} \right) + c_2 R_2\left(G^{iter} - X_i^{iter} \right) \right] \tag{6.5}$$

The constriction factor is defined as,

$$K = \frac{2}{|2 - \phi - \sqrt{\phi^2 - 4\phi}|} \dots \varphi = c_1 + c_2, \varphi > 4 \tag{6.6}$$

PSO with the constriction factor has been found to have improved convergence. Choosing appropriate values for w_v, c_1, and c_2 ensures convergence without the need for velocity clamping. The PSO algorithm (Shi and Eberhart) with only a constriction factor included results in an improved rate of convergence, but sometimes the threshold might not be reached within the specified number of iterations. To overcome this problem, the velocity was clamped to the maximum limit and the performance improved. Therefore, using the constriction factor along with velocity clamping improves the rate of convergence, and the convergence was reached within the specified number of iterations [13]. A typical value for φ is equal to 4.1 (>4), leading to $K = 0.7298$ and since $\varphi = c_1 + c_2$, $c_1 = c_2 = 2.05$ is a good choice for the coefficients.

The discrete binary version of PSO [14] was proposed by Kennedy and Eberhart in 1997. In binary PSO, the components of the vector representing present position, *personal best*, and *global best* are binary in nature, meaning they assume values of either 0 or 1. The position of a particle is represented by a binary number, the length of the number being equal to the dimension d of the search space. The velocity vector components are thresholded to lie in the range [0, 1] using the sigmoidal function defined as $sig(x) = \dfrac{1}{1+e^{-x}}$. The velocity

update equations in binary PSO are same as those in the original PSO. The position update equations are modified as

$$x_{i,j}^{iter+1} = \begin{cases} 1 & if \ r < sig\left(v_{i,j}^{iiter+1}\right) \\ 0 & if \ r \geq sig\left(v_{i,j}^{iter+1}\right) \end{cases} \tag{6.7}$$

where r is a random number that takes on values between [0, 1] with a uniform probability. Equation 6.7 represents the components of the position vector where $j = 1, 2, ..., d$.

When $v_{i,j}^{iter+1} > 10$, $sig(v_{i,j}^{iter+1})$ is saturated at the value of 1. When $v_{i,j}^{iter+1} < -10$, $sig(v_{i,j}^{iter+1})$ is approximately 0. Hence, $v_{i,j}^{iter+1}$ may be clamped to ± 4 or ± 6, as suggested in the literature. This makes $sig(v_{i,j}^{iter+1})$ vary between 0.0180 and 0.9820 for $v_{i,j}^{iter+1}$ clamped to ± 4 and $sig(v_{i,j}^{iter+1})$ varies between 0.0025 and 0.9975 for $v_{i,j}^{iter+1}$ clamped to ± 6. In binary PSO the particles move by flipping bits, since the components of the vectors are binary 1s and 0s. The Hamming distance between two binary vectors is defined as the number of positions in which the elements of the vector differ. The Hamming distance between X_i^{iter+1} and X_i^{iter} is the change in velocity or acceleration of the particle. When there is no change in the bits, the particle is in the same position, whereas if all the bits are flipped the particle moves the farthest distance.

There may be multiple objective functions [15] or there may be a single objective function with multiple constraints. In these cases, the optimum has to be found that is a tradeoff between the multiple functions or constraints. There will be a set of solutions that satisfy the multiple objectives which could be a tradeoff between the multiple conflicting objectives and constraints.

6.5 Summary

The algorithm is stochastic, does not require computation of gradients, and is based on the behavior and movement dynamics of swarms of birds. There are only a few parameters to be controlled, and it is computationally efficient, derivative-free, simple to implement, and applicable to a wide range of problems. Because there is no necessity for calculation of derivatives, computational complexity is reduced. It requires less code size and memory and is quite fast in execution, leading to reduced space and time complexity. Unlike GA where the population changes with every generation or iteration, in PSO the population remains constant throughout the run of the algorithm. The typical population size is 10 to 50, chosen depending on the problem. The uniqueness of PSO lies in the fact that possible solutions (particles) fly through the solution space (hyperspace) accelerating towards better solutions. It is a stochastic algorithm that has a certain amount of randomness built into it. If there are constraints in the problem, in effect, the search space is reduced. The global optimum solution should satisfy all the constraints and the objective(s) of the problem.

The algorithm must strike a balance between exploration and exploitation. Exploration will lead to exploring the search space in new areas whereas exploitation will make the algorithm search in a local (smaller) region intensely. The choice of parameters is crucial for this balance. Smaller values of w_v result in faster convergence. The particle converges usually on the line between the *personal best* and the *global best* positions. As the number of dimensions increases, the time taken for the algorithm to converge also increases. The parameters c_1, c_2, and w_v may be held constant for all the iterations, or c_1, c_2 may be held constant with linearly decreasing inertia weight w_v for the iterations. An initial large inertia weight leads to exploration of the search space, and as the weight decreases, it increases the exploitation abilities of the swarm. Objective functions can be unimodal or multimodal. Multimodal functions require more diversity among the swarm than unimodal functions. Several hybrid algorithms have been proposed where evolutionary operators have been applied with PSO, or other swarm intelligence algorithms have been combined with PSO leading to enhanced performance for certain applications.

References

1. James Kennedy, Russell Eberhart, Particle swarm optimization, *Proceedings of the IEEE International Conference on Neural Networks*, Piscataway, NJ, Vol. IV, pp. 1942–1948. IEEE Press, 1995.
2. Mahamed G. H. Omran, Particle swarm optimization methods for pattern recognition and image processing. Ph.D. thesis. University of Pretoria, Pretoria, 2005.
3. Frans van den Bergh, An analysis of particle swarm optimizers. Ph.D. thesis. University of Pretoria, Pretoria, 2001.
4. E. O. Wilson, *Sociobiology: The New Synthesis*, Cambridge, MA: Belknap Press, pp. 697, 1975.
5. James Blondin, Particle swarm optimization: A tutorial, September 2009. http://cs.armstrong.edu/saad/csci8100/pso_tutorial.pdf
6. James Kennedy, Russell Eberhart, Yuhui Shi, *Swarm Intelligence*, Morgan Kaufmann Publishers, San Francisco, 2001.
7. Reynolds, C. W., Flocks, herds, and schools: A distributed behavioral model, *Computer Graphics*, Vol. 21, No. 4, *SIGGRAPH '87 Conference Proceedings*, pp. 25–34, 1987.
8. Millonas M. M, Swarms, Phase transitions and collective intelligence, In: *Artificial Life III*, C. G. Langton (ed). Reading, MA: Addison Wesley, pp. 417–443, 1993.
9. M. Clerc, J. Kennedy, The particle swarm - Explosion, stability, and convergence in a multidimensional complex space, *IEEE Transactions on Evolutionary Computation*, Vol. 6, No. 1, pp. 58–73, 2002.
10. Y. Shi, R. Eberhart, A modified particle swarm optimizer, *1998 IEEE International Conference on Evolutionary Computation Proceedings, IEEE World Congress on Computational Intelligence*, Anchorage, AK, USA, pp. 69–73, 1998.
11. Dian Palupi Rini, Siti Mariyam Shamsuddin, Siti Sophiyati Yuhaniz, Particle swarm optimization: Technique, system and challenges, *International Journal of Computer Applications*, Vol. 14, No. 1, pp. 19–27, January 2011.
12. M. Clerc, The swarm and the queen: Towards a deterministic and adaptive particle swarm optimization, *Proceedings of the Congress on Evolutionary Computation*, Washington, DC, United States, pp. 1951–1957, July 1999.
13. Russ C. Eberhart, Y. Shi, Comparing inertia weights and constriction factors in particle swarm optimization, *Proceedings of the Congress on Evolutionary Computation*, San Diego, CA, United States, pp. 84–89, 2000.

14. James Kennedy, Russell Eberhart, A discrete binary version of the particle swarm algorithm, *Proceedings of the Conference on Systems, Man and Cybernetics*, Orlando, FL, USA, pp. 4104–4109, 1997.

15. E. Zitzler, M. Laumanns, S. Bleuler, A tutorial on evolutionary multiobjective optimization, In: *Metaheuristics for Multiobjective Optimization*, X. Gandibleux, M. Sevaux, K. Sörensen, V. T'kindt (eds), Lecture Notes in Economics and Mathematical Systems, Vol. 535. Berlin, Heidelberg: Springer, pp. 3–37, 2004.

7

Differential Evolution

7.1 Introduction

Differential evolution (DE) was invented in 1995 by Price and Storn and has been found to be robust in solving global optimization problems [1]. Any good optimization algorithm should converge in finite time and produce the best (optimized) output. The complex engineering problems are stated in terms of objective(s) to be achieved and constraints within which the optimum solution is to be attained. These objective(s) and constraints are expressed as mathematical equations that can be solved using standard numerical techniques in the case of traditional optimization methods. The mathematical function representing the objective(s) to be attained is termed as the objective function that is to be either minimized or maximized. When the function is minimization of an objective, it becomes a cost or error function. When the function is maximization of an objective, it becomes a profit or quality function. The problem to be solved might have constraints that could be incorporated into the formulation of the objective function, or they might be separately expressed as mathematical equations. The objective function could be continuous or discrete, linear or non-linear, differentiable or non-differentiable. The objective function is dependent on a set of variables whose values determine the value of the function (output). In addition, the objective function might be dependent on a set of parameters related to the problem. The appropriate choice of parameters for the problem leads to the global optimum solution [2].

In DE, it is assumed that the fitness or objective function is a cost function that is to be minimized. Differential evolution was proposed to be a stochastic direct search method to find the global minimum, with inherent parallelism to take care of computationally intensive cost functions. DE basically uses a population of vectors whose perturbations can be done independently in the search space which is a polyhedron. Each vertex of the search space is represented by a d-dimensional population vector. When the population vectors need to be transformed or evolved, change in the control variables or parameters might be required. DE has a self-organizing capability to determine perturbations of the population vectors in the search space without having to apply changes in the control variables or parameters. The perturbations are applied for all the population vectors independently. This is in contrast to traditional algorithms that apply pseudorandom parameters with a uniform probability distribution to bring about such perturbations. DE has a lower number of control variables to tune the algorithm. DE has been proven experimentally to have good convergence properties. DE can handle linear or non-linear, differentiable or non-differentiable, unimodal or multimodal functions. DE is a derivative-free search strategy that is suitable for unimodal as well as multimodal functions.

7.2 Differential Evolution

Differential evolution, as the name suggests, is an optimization strategy where evolution of the population members or vectors takes place based on differences between existing vectors of the population [3]. DE starts with an initial population of randomly chosen points in the search space, where each point is represented by a vector. The population size is assumed as N and the dimension of the search space is d. Hence, each vector of the population is d-dimensional and has d number of components. The search space is bounded by the limits on the parameters, and the initial population should be chosen so that it is distributed uniformly (pseudorandom distribution) throughout the search space. If there is a vector already available in the search space (preliminary nominal solution vector), the initial population is generated by adding randomly generated deviations to the available vector. This process creates the initial population of vectors in the search space.

New population members or vectors are generated from perturbations of existing vectors [4]. Two vectors from the existing population are chosen randomly and their difference is taken. This difference vector is multiplied by a weighting factor to create the perturbation vector. The perturbation vector is added to a third randomly chosen population vector, to generate a new vector in the search space. This process of generating new vectors is called *mutation* in DE.

A population vector is randomly chosen in the search space called the *target* vector. The vector obtained as a result of mutation is mixed with the target vector to generate a new vector called the *trial* vector. This process of mixing vectors is called *crossover* in DE. The objective function is evaluated with the *trial* vector and *target* vector and they are compared. The vector with the lower cost value is retained in the next generation of the population. This process of replacing the *target* vector with the *trial* vector if the *trial* vector has lower cost than the *target* vector is called *selection*. This procedure of applying *mutation, crossover*, and *selection* is repeated until all the members of the population have been covered. Every member of the population of size N becomes a *target* vector once during every iteration. The survivors of *mutation, crossover*, and *selection* operations are the next-generation members of the population.

DE starts with an initial population size N. The population consists of d-dimensional vectors; in other words, every vector of the population has length d. The upper and lower bounds for all the parameters involved in the problem are defined. One notable factor in DE is that all parameter values are real numbers and are represented in floating point notation. As already stated, randomly chosen population vectors are mutated by adding the weighted difference of two randomly chosen vectors that are different from the original vector. The weighting factor is also a random number that normally assumes positive values in the range [0, 2]. In addition to mutation, DE also employs crossover operation. In crossover, the trial vector is created by building the vector element by element from the mutant vector and the chosen target vector. The crossover constant c_r controls the number of elements that are copied from the mutant vector. The crossover constant is randomly chosen between [0, 1], and it is compared with another random number $rand(i)$ generated. Depending on whether crossover constant c_r is higher or lower than the random number $rand(i)$ and also considering the index k of the element, the crossover operation is done. If the trial vector created by crossover has lower cost than the corresponding target vector, the trial vector replaces the target vector; otherwise the target vector is copied into the next-generation population. These processes of *mutation, crossover*, and *selection* are repeated in every iteration (generation) until the maximum number of iterations is reached or a termination criterion is met or the optimum solution is attained.

The difference vectors can be both positive as well as negative, and hence the mean of their distribution is zero. Scaling the difference vectors (multiplying by the weighting factor) ensures that the trial vectors are not repeated (no duplication of vectors) and also the search does not get trapped in local minima. As the number of iterations increases, the population vectors cluster towards the global minimum and the difference vectors have a length and direction suitable for local search. Since DE codes all parameters as real numbers with floating point notation, it is easy to use, has efficient memory usage, lower computational complexity, and converges faster. The number of bits to represent the fractional part of the number is the precision, and this must be high in order to represent very small differences. The number of bits allotted for the exponent determines the magnitude of the numbers (very large or very small) that it is possible to represent. Therefore, floating point representation is very much useful in DE since it involves computation of differences which could be very small values. These small changes could have a profound effect on the performance of DE, and hence they should be captured. This is facilitated by the floating-point notation employed.

The initial population has to be chosen that is uniformly distributed within the search space. The initial population not only requires distribution over the entire search space but some probability distribution must be followed in choosing the initial vector elements. Mostly, uniform distribution is followed in choosing the initial population elements, but Gaussian distribution makes convergence faster, sometimes prematurely. Some randomness should be included in the choice of initial parameter values. Clustering the initial population has a detrimental effect on the performance of the algorithm, so distribution of the initial population throughout the bounded search space is necessary. In standard test functions, the limits on the parameters could be specified by the user to determine the boundary of the search space. In real-world applications, the objective function and the parameters are dependent on the problem to be solved. The minimum and maximum values of the parameters also will be problem-dependent, and care must be exercised in fixing such extreme values since they have a great impact on the search for the optimum solution.

In the DE algorithm, the base vectors for calculating weighted differences, the population vector on which mutation is performed, and the target vector on which crossover is done are all chosen randomly. There is a possibility that some of the vectors might be chosen repeatedly while others might not get a chance to contribute to the evolution process. This will have a profound effect on the performance of the DE algorithm. In addition to random selection of base vectors as in the original DE algorithm, the choice could be based on fitness values of the vectors, as proposed in other versions of the DE algorithm. The algorithm finds the optimum solution when it stops at the appropriate condition, called the termination criterion. When the objective(s) and constraints are met, the algorithm stops execution. The optimum value of the objective function or at least the neighborhood of the optimum must be previously known, and when the constraints on the problem have been satisfied the algorithm can terminate. In single-objective optimization, it is easy to identify the optimum point, but in multi-objective optimization, it might not be easy to identify the termination point since it could involve multiple conflicting objectives that might be difficult to satisfy. In most of the problems, the objective function value might not be known so the algorithm stops with a preset maximum number of generations or iterations. The algorithm could also be made to terminate when there is no sizeable improvement in the fitness function value with consecutive, succeeding iterations.

7.2.1 Algorithm

Let the population size be N and let d be the dimension of the population vectors. Let the population vector be represented by

$$X_i^{iter} = \left[x_{i1}^{iter} \; x_{i2}^{iter} \; ... \; x_{id}^{iter} \right], \quad i = 1, 2, ..., N \tag{7.1}$$

where *iter* is the variable for indexing the iterations and the maximum number of iterations is *MaxIter*, which is the maximum number of iterations for the algorithm to run. The initial population of vectors is chosen randomly, and distributed uniformly over the entire search space within the boundaries.

The population evolves in DE by perturbing the existing population vectors. The perturbation is obtained as a difference of two of the existing population vectors (base vectors) multiplied with a weighting factor. The perturbation is added to a third member of the population vectors (different from the base vectors) to create the mutant vector. This operation is called *mutation* in DE, and there is one mutant vector M_i for every target vector X_i. The *mutation* operation can be mathematically described as:

$$M_i^{iter} = X_r^{iter} + w_i . \left(X_j^{iter} - X_k^{iter} \right), \quad i, j, k, r \in \{1, 2, ... N\} \tag{7.2}$$

where N is the total population size, M represents mutation vector, i, j, k, and r are indices for vectors chosen randomly from the population, and w_i is the weighting factor that has a positive value (>0). This implies a minimum population size of four. For every vector X_i of the population there is a mutant vector M_i generated. Therefore, Equation 7.2 is applied for every X_i, $i = 1, 2, ..., N$. The process of generating mutant vector M_i from the population vectors is diagrammatically illustrated in Figure 7.1 for a two-dimensional search space.

The elements of the *mutant* vector are mixed with the elements of a *target* vector chosen from the population, and this operation is called *crossover*. This *crossover* operation creates the *trial* vector. The crossover operator is introduced in DE to increase the diversity of the population. Let the target vector and the corresponding mutant vector in iteration *iter* be given by Equation 7.3.

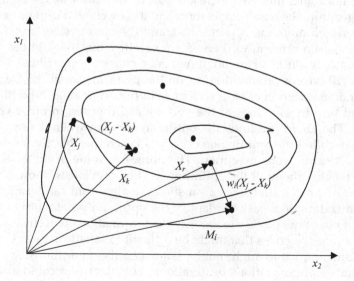

FIGURE 7.1
Mutant vector generation.

$$\text{Target vector}: X_i^{iter} = \{x_{i1}^{iter}, x_{i2}^{iter}, \ldots, x_{id}^{iter}\}$$

$$\text{Mutant vector}: M_i^{iter} = \{m_{i1}^{iter}, m_{i2}^{iter}, \ldots, m_{id}^{iter}\} \tag{7.3}$$

Let the trial vector generated by the *crossover* operation be given by:

$$\text{Trial vector}: Y_i^{iter} = \{y_{i1}^{iter}, y_{i2}^{iter}, \ldots, y_{id}^{iter}\} \tag{7.4}$$

where the elements of the new crossover vector y_i^{iter} are generated as follows.

$$
\begin{aligned}
y_{ik}^{iter} &= m_{ik}^{iter} \quad \text{if} \quad rand(k) \leq c_r \quad \text{or} \quad k = r(i) \quad k = 1, 2, \ldots, d \\
&= x_{ik}^{iter} \quad \text{if} \quad rand(k) > c_r \quad \text{and} \quad k \neq r(i)
\end{aligned}
\tag{7.5}
$$

c_r is defined as crossover constant that takes on values in the interval [0, 1], $rand(k)$ is a random number generator for the element with index k that assumes values in the range [0, 1], and $r(i)$ is a randomly chosen index from the set $\{1, 2, \ldots d\}$. Crossover is illustrated in Figure 7.2.

The objective function is evaluated for the target vector X_i^{iter} chosen and the trial vector Y_i^{iter} generated. This gives the values of $f(X_i^{iter})$ and $f(Y_i^{iter})$. If $f(X_i^{iter})$ is less than $f(Y_i^{iter})$ then, $X_i^{iter+1} = X_i^{iter}$, else $X_i^{iter+1} = Y_i^{iter}$. This is a *selection* operation for the next-generation population vectors. The vectors of the next generation are created by evaluating the objective function on the chosen target and trial vectors and the vector with the lower cost function value is included in the next generation. This process is repeated for all the target vectors of the current generation ($i = 1, 2 \ldots d$) to evolve the population for the next generation. One

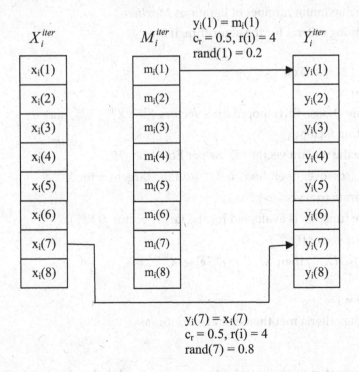

FIGURE 7.2
Illustration of *crossover*.

method of updating the population is performing *mutation, crossover,* and *selection* for all the (target) vectors of the population and storing the new vectors separately. Once these operations are completed for the entire population, the old population is replaced with the new vectors. As a variant of this, as each trial vector is generated and either the target or trial vector is selected (based on the fitness value), the population can be updated adaptively as the iterations proceed. This could possibly lead to faster convergence.

As a *rule of thumb,* the population size for DE should be five to ten times the dimension d of the population vectors, with a minimum population size of four. The weighting factor w can be chosen as 0.5, and depending on the rate of convergence of the algorithm, the weighting factor can be increased or decreased. In general, the weighting factor can be positive with a maximum value of 2. The crossover constant c_r can be chosen approximately close to 1 such as 0.9 or 0.8, but again it can be reduced depending on the rate of convergence. Most of the parameters that are stochastic follow a uniform distribution. DE has been applied and tested on a set of benchmark functions and found to outperform several variants of the annealing and genetic algorithm. The computational complexity is less for DE compared to other evolutionary algorithms. DE has been mainly proposed for minimization functions although it can be applied for function maximization too.

7.2.2 Pseudocode

Initialize the population size N and the dimension of the vectors d

Randomly generate the initial population of vectors in the search space $\{X_i\}$, $i = 1, 2, \ldots, N$

Define the fitness or objective function $f(X)$

Choose the maximum number of iterations *MaxIter*

Define stopping criteria for the algorithm, if any

$iter = 1$

while ($iter \leq MaxIter$) **do**

  for $i = 1$ to N

    Randomly choose three population vectors X_r^{iter}, X_j^{iter}, X_k^{iter}, and weighting factor w_i

    Generate the mutant vector M_i^{iter} as per Equation (7.2)

    Perform crossover operation on M_i^{iter} and the target vector X_i^{iter} to generate trial vector Y_i^{iter}

    Objective function is evaluated for the target vector $f(X_i^{iter})$ and trial vector $f(Y_i^{iter})$

    if $f(Y_i^{iter}) < f(X_i^{iter})$ **then** $X_i^{iter+1} = Y_i^{iter}$ **else** $X_i^{iter+1} = X_i^{iter}$

  end for

  $iter = iter + 1$

  if stopping criteria met **then** exit **else** continue

end while

Population vector with lowest fitness value is global optimum (minimization problem)

Flowchart

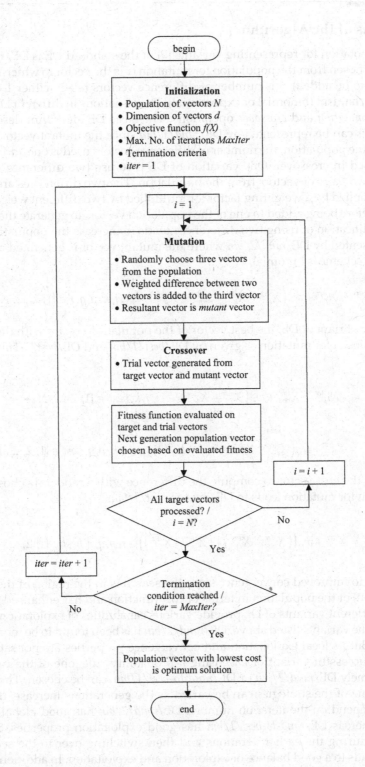

7.3 Variants of the Algorithm

The general notation for representing variants [5] of the standard DE is DE/x/y/z where x is the vector chosen from the population for mutation (x is the vector to which the weighted difference is to be added), y is number of difference vectors taken (either 1 or 2), z is the crossover mechanism (binomial or exponential). The variations in different DE algorithms occur in the *mutation* and *crossover* operations. The basic DE algorithm described in the above sections can be represented by DE/*rand*/1/*bin* since the mutant vector is randomly chosen from the population, the number of difference vectors used is one and the binomial method is used in crossover. One variation of DE is to use two difference vectors computed from two pairs of vectors from the population. The two differences are added, and they are multiplied by a weighting factor (or multiplied by two different weighting factors separately) before being added to one of the population vectors to generate the mutant vector. This modification of using two differences tends to increase the population diversity. This is represented by DE/*rand*/2/*bin* where the mutant vector is generated as below and the crossover scheme is binomial.

$$M_i^{iter} = X_r^{iter} + w_i \cdot \left(X_j^{iter} - X_k^{iter} + X_p^{iter} - X_q^{iter} \right), \quad i,j,k,p,q,r \in \{1, 2, ... N\} \tag{7.6}$$

In yet another variant of DE, the best vector of the population (vector with the lowest cost function) is chosen for mutation as given by DE/*best*/1/*bin* and DE/*best*/2/*bin* respectively as below:

$$M_i^{iter} = X_{best}^{iter} + w_i \cdot \left(X_j^{iter} - X_k^{iter} \right), \quad i,j,k,best \in \{1, 2, ... N\} \tag{7.7}$$

$$M_i^{iter} = X_{best}^{iter} + w_i \cdot \left(X_j^{iter} - X_k^{iter} + X_p^{iter} - X_q^{iter} \right), \quad i,j,k,p,q,best \in \{1, 2, ... N\} \tag{7.8}$$

Finally, using the best vector to compute the difference with a randomly chosen vector of the population for mutation leads to DE/*rand-to-best*/1/*bin*:

$$M_i^{iter} = X_r^{iter} + w_i \cdot \left(\left(X_{best}^{iter} - X_r^{iter} \right) + \left(X_p^{iter} - X_q^{iter} \right) \right), \quad i,p,q,r,best \in \{1, 2, ... N\} \tag{7.9}$$

This can lead to improved convergence since the search is in the region of the best vector (feasible region) of the population in terms of cost function and has enhanced local exploitation. The different variants of DE provide various capabilities of exploration and exploitation. Of all the variants listed above, DE/*rand*/2/*bin* has been found to be more successful than others. But its local exploitation and convergence properties are not satisfactory. In yet another successful variant of DE, using a probability rule, one of the two mutation strategies, namely DE/*rand*/2/*bin* or DE/*rand-to-best*/1/*bin*, can be chosen. The probability of choosing one of the strategies can be varied as the generations increase, that is, it can be made to depend on the iteration number. DE/*rand*/2/*bin* has good global exploration properties whereas DE/*rand-to-best*/1/*bin* has good exploitation properties. So choosing the first one during the earlier iterations and then switching over to the second one at later stages leads to a good balance of exploration and exploitation. In addition, during the middle phase of iterations, either one of the strategies can be used. Different probability rules can be applied in the selection of the two strategies. Mutation is responsible for introducing diversity and hence global search capability of the algorithm.

Similarly, two crossover methods, *exponential* and *binomial*, are used in DE. The crossover strategy described above is the *binomial* method since the trial vector gets its components from the mutant vector following a binomial distribution. In *exponential crossover*, two integers are randomly chosen; one is the index for the starting point for crossover, and the other is the number of components of the trial vector that it gets from the mutant vector. The crossover constant determines the components of the trial vector, that is, how many components of the trial vector are inherited from the mutant vector. Smaller values of the crossover constant lead to lesser diversity and slower convergence whereas larger values of the crossover constant increase the diversity as well as the convergence rate. In the earlier iterations, the crossover constant has to be a small value, typically less than 0.2, in order to have good exploratory capabilities, and as the iterations advance, the crossover constant can be increased above 0.8. This adaptive strategy balances exploration and exploitation. During the iterations, the population vectors could move out of the boundaries of the search space but it is in-built in the DE algorithm not to exceed the boundaries. This could cause the population vectors to lie on the boundaries of the search space, thus reducing the diversity. This could be overcome by using a strategy to move the population away from the boundaries of the search space when the vectors move to the boundary during the iterations. The appropriate choice of the two parameters, weighting factor and the crossover constant, are very important in the performance of the DE algorithm. These parameters could be made self-adaptive so that the algorithm converges faster and does not get trapped in local optima.

In another variant of the DE algorithm, both uniform and Gaussian distributions are used in selecting the scale factor and crossover constant for unconstrained optimization problems [6]. This improves the diversity of the population. Hybrid DE algorithms have been proposed by combining DE with other optimization strategies to improve the performance. An archive with high-quality solutions can be included during the process to improve the quality of the population. In the literature, DE has been combined with *k*-means clustering to improve the performance for unconstrained optimization problems. For problems requiring extensive computations, the DE has been combined with a *k* nearest neighbor (kNN) approach to improve the performance. This hybrid variant uses a predictor that predicts a good approximation to the actual space and is efficient. In *modified DE* (MDE) there are four variants to the original DE algorithm. MDE uses an external archive to store good candidate populations that are used during the run of the algorithm. Two mutation strategies DE/*rand*/1/*bin* and DE/*best*/1/*bin* are applied and either of them chosen using a probability [7]. The scale factor and crossover constant are modified adaptively with the iterations. Gaussian distribution is used for modifying the scale factor, and uniform distribution is used for modifying the crossover constant. Out of the N members of the population, all of them are updated according to the update equations of the algorithm except one member whose position is the average of all other members $(N - 1)$. This is the central solution that is an alternative to the optimum solution. In novel MDE (NMDE) the scale factor and crossover rate are adaptively modified and each solution has its own parameters as scale factor and crossover rate instead of their being same for the entire population [8]. In the original DE algorithm, the parameters are fixed throughout the run of the algorithm once they are chosen. In NMDE the parameters are adaptive so that the algorithm can come out of any local optimum. This NMDE in combination with a penalty function method is suitable for solving constrained optimization problems. DE has also been modified to solve multi-objective and combinatorial optimization problems where the search space is not continuous. Some of the hybrids with promising performance are biogeography-based optimization (DE-BBO), estimation of distribution algorithm (DE-EDA), fittest individual refinement (DE-FIR), DE-barebones particle swarm optimization (PSO), and neighborhood search DE, to name a few variants. These hybrid variants of DE balance exploration and exploitation and converge faster, leading to better solutions than either of the algorithms run separately. The main contributions of the DE variants are in the mutation strategy and choice of crossover parameters.

7.4 Summary

DE is a global optimization technique that is easy to use, reliable, fast, and simple to implement. DE has proven to be a promising approach to solve problems that are non-linear and non-differentiable. DE can efficiently handle unimodal as well as multimodal objective functions that are computationally intensive. It has few control variables, is robust, converges consistently and faster, and has been found to be more computationally efficient than other classical optimization methods. It has an implicit parallelism that increases its rate of convergence. DE is a direct search method that is easy to implement and gives good results compared to other evolutionary algorithms. DE has been proved to be efficient in solving engineering design problems and real-life applications in diverse fields. DE operates on continuous spaces and uses distance and direction information in the form of difference vectors in conducting the search for the optimum solution. DE has a self-organizing capability that makes it remarkably different from other optimization techniques.

References

1. Kenneth Price, Rainer Storn, Differential evolution – A simple and efficient adaptive scheme for global optimization over continuous spaces, Technical Report TR – 95 – 012, International Computer Science Institute, Berkeley, CA, United States, 1995.
2. Kenneth V. Price, Genetic annealing algorithm, *Dr. Dobb's Journal*, pp. 127 – 132, October 1994.
3. Rainer Storn, Kenneth Price, Differential evolution – A simple and efficient heuristic for global optimization over continuous spaces, *Journal of Global Optimization*, Vol. 11, pp. 341–359, Kluwer Academic Publishers, 1997.
4. Kenneth V. Price, Rainer M. Storn, Jouni A Lampinen, *Differential Evolution – A Practical Approach to Global Optimization*, Springer Natural Computing Series, Springer-Verlag, Berlin, Heidelberg, 2005.
5. Swagatam Das, Sunkha Subhra Mullick, P. N. Suganthan, Recent advances in differential evolution – An updated survey, *Swarm and Evolutionary Computation*, Vol. 27, pp. 1–30, 2016.
6. Dexuan Xou, Jianhua Wu, Liqun Gao, Steven Li, A modified differential evolution algorithm for unconstrained optimization problems, *Neurocomputing*, Vol. 120, pp. 469–481, 2013.
7. Xiangtao Li, Minghao Yin, Modified differential evolution with self-adaptive parameters method, *Journal of Combinatorial Optimization*, Vol. 31, pp. 546–576, 2016.
8. Dexuan Xou, Haikuan Liu, Liqun Gao, Steven Li, A novel modified differential evolution algorithm for constrained optimization problems, *Computers and Mathematics with Applications*, Vol. 61, pp. 1608–1623, 2011.

8

Ant Colony Optimization

8.1 Introduction

Swarm intelligence has been the inspiration behind the development of a class of nature-inspired optimization algorithms that are different from the traditional methods. These optimization techniques are unconventional and have been found to be successful in solving a diverse range of real-life problems. The traditional optimization algorithms are suitable for continuous functions that require the computation of derivatives. The nature-inspired optimization algorithms can be applied for continuous and discrete as well as mixed-variable problems, and they do not require the computation of derivatives. They are mostly search algorithms that use a population of agents to search in parallel for the optimum, thus saving time. Metaheuristics is an important component of such algorithms since approximations greatly simplify the process in arriving at the optimum solution for the problem.

Ant colony optimization (ACO) is one such swarm intelligence-based metaheuristic algorithm and was proposed by Marco Dorigo and Gianni Di_Caro in 1999 [1]. ACO was developed with the inspiration of the foraging behavior of ant colonies [2]. The collective intelligence of a swarm of ants has been used to solve intractable problems that are NP-hard for conventional algorithms. Inspiration from the study of the swarm behavior of ant colonies has led to the development of the ACO algorithm. Ants always tend to follow the shortest path from their nest to a food source, and this technique has been inculcated into the ACO algorithm to find the shortest path in graph-based search. This class of problems where the optimum solution is the shortest route in an interconnected graph or a network has been successfully solved by the ACO algorithm. Since the first ant-based algorithm was proposed by Marco Dorigo in 1992, several variants of the algorithm have been proposed by researchers working in the field, and they have been successfully applied to problems that involve a tedious searching process with a lot of parameters. Ants provide inspiration, and knowledge derived from their behavior can provide solutions to discrete combinatorial optimization problems. ACO can be easily extended to all problems under the discrete combinatorial optimization category with minor variations. The algorithm is population-based; hence it is suitable for parallel search, thus reducing the time complexity. The characteristics of ant colonies and the ACO algorithm with its variants have been discussed in the following sections.

8.2 Ant Colony Characteristics

The ants are well-organized in colonies, and they exhibit collective intelligence. There are thousands of species of ants throughout the world such as fire ants, army ants, black ants, red ants, carpenter ants, pharaoh ants, field ants, and so on. Some of the species of ants build nests indoors whereas others live in fields or outdoor grounds. In an ant nest there is a hierarchy of queen and workers. There could be one queen ant or more than one queen ant in a nest. The worker ants (both young and old ants share the duties) take care of the regular duties of the nest such as cleanliness, defense of the colony, looking after larvae, foraging, etc. At times, the duties are re-allotted or shared according to the prevailing environmental conditions, making the ant colony flexible and versatile. Figure 8.1 shows Eastern carpenter ants (*Camponotus pennsylvanicus*) in Ontario, Canada. A typical ant hill and ant tracks at the Oxley Wild Rivers National Park, New South Wales, are shown in Figure 8.2.

Ants exhibit collective intelligence and accomplish tasks together as a swarm rather than individually. Together the ants forage for food in different directions, and they coordinate with each other for the benefit of the entire swarm. When there are multiple paths between the ant nest and the food source they tend to find the shortest path between the nest and food source and this is followed by all the ants. Ants generally find food and store it in their nests for rainy days. The decisions on how many ants are allotted for foraging, the quantity of food to store, and food distribution are taken by the swarm collectively. The ants execute such duties efficiently, and this behavior has been modeled into the ACO algorithm. Some of the ants have wings, enabling them to fly looking for mates, while others that do not have wings move over surfaces while performing their duties, and they are mostly worker ants. Figure 8.3 shows a swarm of harvester ants transporting food (seeds) into their nest.

Ants deposit a chemical called pheromone on the path they travel which can be sensed and followed by other ants. Ants trace a path between their nest and source of food by depositing pheromone [3]. When an ant moves on the path it secretes pheromone, and higher pheromone deposits increase the probability of more ants following the path. Figure 8.4 shows a group of ants following pheromone trails. Ants use the scent of the

FIGURE 8.1
Eastern black carpenter ants. (Author: Ryan Hodnett – own work, CC BY-SA 4.0. https://creativecommons.org/licenses/by-sa/4.0/deed.en.)

FIGURE 8.2
Ant hill and ant tracks. (By Cgoodwin – own work, CC BY-SA 3.0. https://creativecommons.org/licenses/by-sa/3.0/deed.en.)

FIGURE 8.3
Harvester ants carrying seeds. (Author: Donkey Shot – own work, CC BY-SA 3.0. https://creativecommons.org/licenses/by-sa/3.0/deed.en.)

FIGURE 8.4
Foraging ants. (Photographer: Kris Mikael Krister (imported from 500 px), CC BY-SA 3.0. https://creativecomm
ons.org/licenses/by/3.0/deed.en.)

chemical pheromone and the sun to find their way back to the nest. Ants also use scents to identify members of their own nest, and if an ant does not comply with this, it is rejected from the nest.

Stigmergy is the technique of indirect communication between agents or insects by means of changes in local environment. It is a non-symbolic form of local communication where the agents make a change in the environment by leaving a trace that can be accessed by other agents to perform some action. This reinforces the activity by multiple agents, leading to good systematic outcomes. The communication is only within the neighborhood of the agent that is leaving a trace by releasing some chemical. This leads to self-organized behavior and collective intelligence among the group of agents that can accomplish complex tasks with simple collaboration and without much elaborate planning and control.

Ants communicate by indirect means through the secretion of a chemical called pheromone. The pheromone evaporates after some time and it is limited to the local environment of the ant. This is stigmergy exhibited by ants. Ants leave their nest looking for food. When they move on a trail, they deposit the chemical pheromone. When the ant returns to the nest with food, it deposits more pheromone, thus reinforcing the trail. This chemical is detected by other ants, and they follow the trail to find the food source. This in turn increases the number of ants on the trail and hence the pheromone deposit. If the ant returns to the nest without food (ant is not able to locate a food source), it will not deposit pheromone on the way back to the nest. Pheromones gradually evaporate with time. If there is no pheromone on a path, the ants diverge and forage for food in other directions or areas. The pheromone concentration will be higher on the paths that link an ant nest to a food source whereas it will be lower on other paths taken by ants during foraging. This is indirect communication and sharing of information by ant colonies. This communication is limited to neighborhoods of the ant nests. The various ant colonies also have differences in their pheromones. This ensures that however long a distance an ant travels, it gets back to its nest safely.

When there are multiple paths between ant nest and food source, ants tend to take the shortest path between their nest and food [4]. When the pheromone levels are higher on one path the ants tend to follow that path rather than the ones that have lower pheromone levels. This makes the ants efficient in transporting food to their nests that have been discovered by other ants belonging to their own nest. This behavior of ants can be studied and easily understood by the illustration in Figure 8.5a and b. Initially the ants randomly follow any one of the paths to the food, depositing pheromone on the way. The ants from the nest go through both paths with random fluctuations in ant population and pheromone concentration between the two paths. Finally, after some time, most of the ants follow only one out of the two paths to the food source. Once pheromone concentration starts to increase on one path, it is further reinforced by more and more ants following that path and hence increasing the pheromone concentration further.

Figure 8.5 shows an ant nest and a source of food connected by two paths. In Figure 8.5a the two paths (path 1 and path 2) connecting the ant nest and food source are of equal length. The ants leave their nest looking for food by following one of the two paths available, depositing pheromone on the way. Initially, the ants randomly chose one of the two paths, leading to random fluctuations in the quantity of the pheromone on the paths. After some time, the ants settle down in either path 1 or path 2 and collect food from the source and bring it to the nest. Most of the ants have settled on path 1 except for one or two, as shown in Figure 8.5a. In Figure 8.5b the two paths connecting the ant nest to the food source are of unequal length (path 1 is shorter than path 2). Since path 1 is shorter than path 2, the ants following path 1 will reach the food source earlier than the ants following path 2. This again leads to increased concentration of pheromones on path 1, and it is further reinforced by more and more ants following them. Finally, all the ants follow the shorter path 1 from the nest to the food source. The difference between Figure 8.5a and b is that in the second case, the random fluctuations in the number of ants following the two routes and hence the pheromone concentration are reduced compared to the first one.

The ants choose one of the two paths with a probability that depends on the pheromone concentration. The higher the pheromone concentration, the higher the probability of an ant choosing that path. Initially an ant that leaves the nest chooses one of the two paths randomly and deposits pheromone on the trail it follows. When it reaches the food source, it takes food and, on the way back, lays more pheromone on the trail. When the ant is taking the shorter path it will deposit pheromones faster than on the longer path. When the path length is shorter, the pheromone concentration is higher whereas if the path length is longer, the pheromone concentration is lower.

This behavior of ants can be incorporated into the ACO algorithm to make it adaptive.

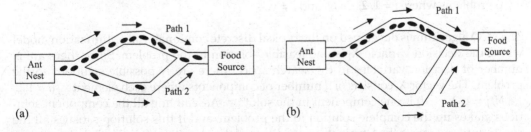

FIGURE 8.5
(a) Two paths of equal length. (b) Two paths of unequal length.

Ants behave in such a manner that their interaction is for the benefit of the entire colony rather than their own individual benefit. The scent of pheromones is sensed by the ants, and they tend to follow the stronger smell of pheromone if they have a choice of more than one path. When an ant finds a food source the quality and quantity of the food located determines the quantity of pheromone deposited by the ant on the way back to the nest after collecting some food. This foraging behavior and interactions between ants have inspired researchers to develop algorithms that mimic their behavior to solve NP-hard problems such as the traveling salesman problem.

8.3 Ant Colony Optimization

Ant colony optimization is a metaheuristic algorithm that was first proposed by Marco Dorigo in the 1990s based on the foraging behavior of ant colonies. In general, the optimization algorithms are either complete or approximate. Complete algorithms find the global optimum solution to the problem either within fixed time limits or might exceed the time bound, whereas approximate algorithms find a good approximation to the optimum solution within the fixed time limits. ACO belongs to the second category of approximate algorithms that find a good (could be the optimum) solution to the problem in a finite reasonable time within bounds. The ACO algorithm was mainly proposed for solving discrete combinatorial optimization problems [5], but it could be modified and extended to continuous and mixed-variable problems. ACO is an iterative algorithm where populations of ants build solutions in every iteration until the stopping criterion is attained.

The ACO algorithm is modeled on the famous traveling salesman problem (TSP) which is a classical discrete combinatorial optimization problem. A typical classical discrete combinatorial optimization problem encompasses the following set of entities:

- Search space S consisting of a finite set of decision variables represented by the d-dimensional vector $X_i = [x_{i1}, x_{i2}, , , , , , , , x_{id}]$, $i = 1, 2, ..., N$. The members of the population are the decision variables, each being a vector of length d.
- The objective or fitness function of the decision variables $f(X_i)$ that is to be either maximized or minimized and represents the quality of the solution obtained.
- Set of equality or inequality constraints defined over the set of decision variables.
- Any solution is called a feasible solution if $f(X_i)$ satisfies all the constraints imposed by the problem. The global optimum solution is the one for which $f(X_g) < f(X_i)$ for minimization problems or $f(X_g) > f(X_i)$ in case of maximization problems where $i = 1, 2, ..., N$, and $i \neq g$.

The ACO algorithm is developed on the typical discrete combinatorial optimization model where the decision variable X_i is one feasible solution to the problem. Since there are N number of decision variables in the search space, there are N possible solutions to the problem. The vector X_i consists of d number of components where each element x_{ij} ($i = 1, 2, ..., N$, $j = 1, 2, ..., d$) is one component in the solution. Assembling all the component solutions makes up the complete solution to the problem, and if this solution satisfies all the constraints and meets the terminating condition of the algorithm, it is the global optimum solution. The ACO algorithm can be developed and easily understood with the classical traveling salesman problem (TSP) which is considered to be NP-hard and intractable and is described below.

8.3.1 Traveling Salesman Problem

Problem Statement: Given a set of cities and paths between the cities (all cities are interconnected), a salesman is to visit each of these cities once and only once by traversing the paths (without retracing) and returning to the starting place so that the total length of the paths traveled and hence the cost are minimum. This is called the *Hamiltonian tour*.

This problem can be described appropriately as a graph G consisting of a set of nodes or vertices representing cities and a set of edges representing paths connecting the cities $G = \{V, E\}$ as shown in Figure 8.6. Let $V = [v_1, v_2, v_3, v_4]$ and $E = [e_{12}, e_{13}, e_{14}, e_{23}, e_{24}, e_{34}]$. There is a cost (proportional to length) associated with every edge represented by $C = [c_{12}, c_{13}, c_{14}, c_{23}, c_{24}, c_{34}]$. Initially, the tour starts from any randomly chosen node of the graph. Let the starting node be v_2. From v_2 there are three unvisited cities v_1, v_3, v_4. The edges and their costs from v_2 to each of these three cities is examined and the edge with the least cost is chosen. Let the edge be v_3 and the path is e_{23} with associated cost c_{23}. From the city v_3 two more cities are to be visited v_1, v_4. The edges e_{13} and e_{34} are examined and let e_{13} be chosen. Then the third city to be visited by the salesman is v_1 and the total cost so far is $C_t = c_{23} + c_{13}$. From the city v_1 the only city not visited is v_4 and the edge connecting them is e_{14} and the cost is c_{14}. From v_4 the salesman has to go back to city v_2 to complete the *Hamiltonian* cycle and the only path to take is e_{24}. Thus the total cost associated with the tour is $C_t = c_{23} + c_{13} + c_{14} + c_{24}$. This is a greedy algorithm which chooses the shortest path or the path with the least cost at every node in each iteration. This case study of the TSP which is a discrete combinatorial optimization problem has been used for developing the ACO algorithm.

The ACO algorithm uses a population of ants to solve the discrete combinatorial optimization problem [6]. The ants build solutions iteratively and deposit pheromones on paths already visited to communicate the quality of the solution found to the other ants indirectly. The population of ants builds a set of feasible solutions to the problem, and the one with the least cost or the best fitness value that satisfies the constraints of the problem is the global optimum solution. The ACO algorithm is modeled as obtaining the complete optimum solution by assembling various solution components. The set of solution components has to be finite, and a pheromone value is associated with each component. The ACO algorithm applied to the TSP [7] is discussed as follows. The graph $G = \{V, E\}$ consists of a set of vertices representing cities and a set of edges connecting the cities. In general, $V = [v_1, v_2, \ldots\ldots, v_N]$ and $E = [e_{12}, e_{13}, \ldots e_{1N}, e_{23}, e_{24}, \ldots e_{2N}, \ldots e_{(N-1)N}]$ where N is the total number of cities. Every edge has a cost or weight associated with it that is proportional to the length of the edge. The problem is to find the *Hamiltonian tour* (global optimum solution)

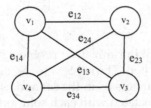

FIGURE 8.6
Graph $G = \{V, E\}$.

by visiting all the cities once and only once (no retracing), and the constraint is that the tour should involve the shortest distance or the least cost.

The search space S consists of all possible tours (solutions) in the graph, randomly starting from any node. Let the set of all possible solutions be $\{X_i\}$, $i = 1, 2, ..., n$. The objective function associated with the solution X_i that is assembled from a set of component solutions is $f(X_i)$. The solution $f(X_i)$ is the sum of the costs (or weight or length) associated with each of the edges that is part of the Hamiltonian cycle. These individual costs associated with the edges are the solution components. One of the ways of representing X_i is as a vector consisting of edges that are part of the solution as evaluated by the function $f(X_i)$. Another way of representing X_i is as a binary vector with length equal to the number of edges in the graph, with a 1 representing inclusion of the edge and a 0 representing an edge not included as part of the solution.

Let e_{ij} represent the edge between nodes (vertices) v_i and v_j and τ_{ij} be the pheromone value associated with the edge e_{ij}. The ant randomly chooses one node (say node v_r) as the starting node. The ant builds the tour by visiting each node once, and it memorizes the nodes already visited. The memory of the ant is denoted as M. Finally it returns to the starting node. The probability of an ant choosing edge e_{ij} is given by

$$p(e_{ij}) = \frac{\tau_{ij}}{\sum\limits_{k=1}^{d} \tau_{ik}} \quad v_k \notin M \tag{8.1}$$

where d is the total number of components in the solution vector, and v_k represents all the nodes not yet visited by the ant and hence is not stored in memory M. Once the solution is constructed, pheromone evaporation is modeled as follows:

$$\tau_{ij} = (1 - r_p)\tau_{ij} \quad i, j \in M \tag{8.2}$$

where r_p is the rate of pheromone evaporation that can take on values in the interval [0, 1]. Here it is assumed that the ants return on the path traced. When the ants return they deposit more pheromone based on the quality of the solution found. The equation modeling this additional pheromone deposit is:

$$\tau_{ij} = \tau_{ij} + \frac{q}{f(X_i)} \tag{8.3}$$

where q is a constant (typical value is 1) and $f(X_i)$ is the cost function (objective function) associated with the solution X_i found by the ant. The number of ants (number of solutions) in each iteration is given by N_A. The iterations are repeated until the termination criteria is attained which is the optimum solution to the problem that satisfies all the constraints.

8.3.2 Algorithm

Generally, for any combinatorial optimization problem, the objective function and the constraints have to be clearly stated. The set $S_C = [s_1, s_2, ...s_C]$ of solution components has to be identified for choosing elements from the set and assembling into a complete solution. The pheromone values associated with each solution component have to be chosen. This involves determining the probability distribution of the pheromone updates, both for pheromone deposit and for pheromone evaporation. This is the pheromone model of the algorithm. Its importance is due to the fact that solution components with higher pheromone values are in the region of the search space where high-quality solutions can be found.

Initially the vector X_i representing the *ith* solution is empty. It is built up iteratively by appending the different component solutions such as $X_i = [x_{i1}, x_{i2}, ...x_{id}]$ where the number of components in the complete solution is d (dimension of the solution vector). Each component x_{ij} is taken from the set S_C, which means $x_{ij} = s_k$, where $k = 1, 2, ..., C$. At the end of building, this solution vector X_i is of length d. In the TSP, the solution components are the edges that are included in the tour. The choice of the component s_k is made by a probabilistic transition model. The pheromone update is done by the ant. Pheromone evaporation prevents convergence towards local optima. Pheromone increase makes the search move in the direction of good quality solutions in the search space. The update also depends on the quality of the solution obtained so far as evaluated by the fitness function. This is included as part of the pheromone update equation.

Ants search for solutions with the least cost in terms of length traveled. Ants memorize their findings (nodes visited), the path taken, and this helps in retracing the path back to the nest, as well as indirectly communicating to the other ants in the colony about the food find. This indirect communication called stigmergy is done by laying pheromone trails. Any ant can trace a path in its neighborhood based on some criteria or in a random manner. The ants from the colony start moving randomly in the neighborhood, and in our model, they move from node to node following edges selected with some transition probability. Every ant changes state in this manner, thus building complete solutions from solution components (each edge selected is a solution component) incrementally. If the termination condition is met even for one ant at the least, the algorithm (search) stops.

The probabilistic transition rule is based on the past history of the moves made by the ant, the fitness value of the solution built so far, the pheromone values existing on the trails, constraints of the problem, and finally some random component (heuristics). Pheromone deposit is done by the ants both during the movement from nest to food as well as during the retrace back to the nest. In the TSP, the ant lays pheromone when it is in transition from one node to another looking for the optimum solution as well as when the ant is retracing its steps to the start node. Good solutions, that is, solutions which are either the global optimum or a close approximation to the global optimum, are attained mainly because of the stigmergy among the ants – their collective behavior and indirect communication. This makes them adaptive to the problem and the environment. Pheromone evaporation favors forgetting and leads to exploration of new regions of search space. If the pheromone evaporation is not modeled into the algorithm it may lead to faster convergence to local optimum.

There is a third, optional component in the ACO algorithm referred to as *Daemon actions*. These actions are for the collective behavior of the entire ant colony and not for single ant. It could be the collection of information about the entire colony of ants or a collective decision about depositing or evaporation of pheromones. The decision to deposit additional pheromones (offline pheromone update) on the shortest paths found so far could be taken for the entire colony of ants. Every ant retraces its path through the graph back to the starting node and deposits additional pheromone on paths with good fitness values. The ant dies after constructing a feasible solution.

The TSP has been chosen as the case study for the ACO algorithms because TSP is one of the classical NP-hard problems which the traditional optimization algorithms cannot solve in finite time. The TSP involves constructing a graph and building solutions incrementally which can be effectively done by ACO. The ACO algorithm easily fits into the

TSP problem and other similar problems like network routing, job scheduling, etc. TSP is one of the earliest discrete combinatorial optimization problems that is easy to understand and comprehend and much researched in the literature. An ant-routing table can be maintained in memory that stores the nodes visited, the edges connected to it, their cost associated, the pheromone concentrations, and the nodes yet to be visited, etc. In each iteration N_A number of ants are employed for the search until the stopping criterion is met. The global pheromone update step is optional. Based on the quality of the solutions, the pheromone concentration may be increased from a global perspective.

An ant evaluates the quality of its tour and if it is good then more pheromones are deposited on those edges to guide the other ants in their search. The pheromone information is exchanged during the tour as a direct indication of the experience of the ant. The memory of an ant (routing table of the algorithm) stores the cities already visited by the ant, cities yet to be visited, edges on which the ant has traveled, the length of the edges, pheromone deposited on the edges by the ant, pheromone evaporation on the edges, the fitness value of the solution obtained so far, etc. In each iteration the number of ants is kept constant. In each iteration the population of ants builds possible solutions by choosing the next vertex or node in the graph based on the pheromone level of the edges. The constraint is that each node should be visited only once. The nodes are selected by the ants with a probability that is associated with the pheromone level of the edge connecting the current node to the next possible nodes (not yet visited). Ants move through the search space looking for solutions collectively. This leads to an exhaustive search and finally the optimum solution is found. Communication between ants is by means of the chemical pheromone in the nearby environment. The algorithm is population-based and iterative, and solutions are improved with every succeeding iteration. The quality of the solutions is evaluated based on the fitness function values for every plausible solution. The quantity of pheromone on a path depends on its length and the number of ants following the trail. The pheromone gets evaporated with time, and shorter paths have more concentration of pheromone compared to longer paths. Pheromone evaporation reduces the convergence of the algorithm to local optimum.

8.3.3 Pseudocode

begin
    Initialization
        Solution components S_C, Number of Ants N_A
        Fitness function $f(X_i)$
        Pheromone update model (transition probability, deposit, evaporation)
        Termination criteria
    while (termination criteria not met) **do**
    Construct Ant Solutions
        Every Ant builds up a solution vector X_i from the component set S_C until the vector length is d (fixed for the problem)
        Component solutions are chosen based on the transition probability
    Pheromone Update
        Pheromone update: evaporation and increase in concentrations done based on the mathematical model

    Daemon Actions \<optional\>
      end while

end

Ant with best solution is the global optimum

Flowchart

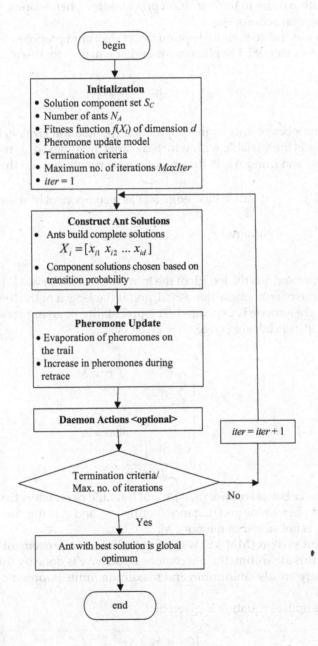

8.4 Variants of the Algorithm

The behavior of ant swarms is primarily to find a path in a search space and deposit pheromones which can be traced by other ants. The path is one which connects the nest of an ant
to a source of food. Therefore, the ACO algorithm and its variants are modeled on a set of
nodes and edges connecting the nodes in the search space. The ant finds the shortest path
through the network of nodes and edges, and that is the optimum solution. The variants
of the ACO algorithm differ in the transition probabilities, pheromone evaporation, pheromone deposit, daemon actions, etc.

The **ant system** was the first original optimization algorithm proposed by Dorigo based on
the behavior of ant swarms [8]. The pheromone update equation for this algorithm is given by:

$$\tau_{ij} = (1 - r_p)\tau_{ij} + \sum_{k=1}^{N_A} \Delta\tau_{ij}^k \tag{8.4}$$

where N_A is the number of ants in the population (population size), r_p is the pheromone
evaporation rate, k is the variable with which the ants are indexed, e_{ij} represents the edge
connecting nodes i and j, and $\Delta\tau_{ij}$ is the pheromone concentration on the edge e_{ij}.

$$\Delta\tau_{ij}^k = \begin{cases} \dfrac{q}{l_k} & \text{if ant k has edge } (e_{ij}) \text{ as a component of its solution} \\ 0 & \text{otherwise} \end{cases} \tag{8.5}$$

where q is a constant and l_k is the length of the tour constructed by ant k. If the length of the
tour is shorter, more pheromone is deposited, and if the length of the tour is longer then a
lower amount of pheromone is deposited. The probability of an ant traveling from city v_i
to city v_j (probability of choosing edge e_{ij}) is given by:

$$p^k(e_{ij}) = \begin{cases} \dfrac{\tau_{ij}^a \lambda_{ij}^b}{\sum\limits_{r=1}^{d} \tau_{ir}^a \lambda_{ir}^b} & \text{if } v_r \notin M \\ 0 & \text{otherwise} \end{cases} \tag{8.6}$$

$$\lambda_{ij} = \frac{1}{dist(i,j)}$$

$dist(i, j)$ is the distance between cities (v_i, v_j), a and b are the parameters that control the relative importance of pheromone and distance (heuristics), and v_r is the city not yet visited by
ant k and hence it is not stored in memory M.

The **max-min ant system (MMAS)** was proposed as an improvement over the original
ant system [9]. In this algorithm, the pheromone update $\Delta\tau$ is done by the best ant instead
of all the ants. There are also minimum and maximum limits imposed on the pheromone
values (bounded).

The pheromone update equation is given by:

$$\tau_{ij} = [(1 - r_p)\tau_{ij} + \Delta\tau_{ij}^{best}]_L^U \tag{8.7}$$

The lower (L) and upper bounds (U) on the pheromone values as given in the above equation depend on the problem. The operator $\left[y \right]_L^U$ is defined as

$$[y]_L^U = \begin{cases} U & \text{if } y > U \\ L & \text{if } y < L \\ y & \text{otherwise} \end{cases} \tag{8.8}$$

$$\Delta \tau_{ij}^{best} = \frac{1}{l_{best}} \quad \text{if } (e_{ij}) \text{ belongs to the best tour}$$

$$= 0 \quad \text{otherwise} \tag{8.9}$$

l_{best} is the length of the tour of the best ant. This l_{best} could be either the best tour in the current iteration, or the best tour in all the iterations completed so far, or a combination of the best in the current iteration and the best attained till now from the start of the iterations.

The **ant colony system** is yet another variation of the ACO algorithm [10] wherein there is a local pheromone update performed by all the ants in addition to the pheromone update at the end of the construction. The local pheromone update is done by all the ants to the last edge traversed at every construction step. The modified pheromone update equation is:

$$\tau_{ij} = (1 - r_p)\tau_{ij} + r_p \tau_0 \tag{8.10}$$

where τ_0 is the initial value of the pheromone, $r_p \in [0,1]$ is the decay coefficient. The pheromone concentration is reduced on the already traversed paths so that subsequent ants do not follow that path and the search gets diversified. This will make ants traverse edges not previously traveled by other ants and produce different solutions.

The pheromone update at the end of the construction process is done by only one ant, either the best in the current iteration or the best in all the iterations so far. The equation for this pheromone update is:

$$\tau_{ij} = \begin{cases} (1 - r_p)\tau_{ij} + r_p \Delta \tau_{ij} & \text{if } (e_{ij}) \text{ belongs to the best tour} \\ \tau_{ij} & \text{otherwise} \end{cases} \tag{8.11}$$

where $\Delta \tau_{i,}^{best} = \frac{1}{l_{best}}$ and l_{best} is either the best in the current iteration or best in all the iterations so far. In this ACS algorithm, the decision rule used by the ants is the pseudorandom proportional rule. The rule states that the probability with which an ant moves from node v_i to node v_j depends on a random variable q that is uniformly distributed in the interval [0, 1] and another parameter q_0. If $q \le q_0$ then $j = \arg\max\{\tau_{ir}\lambda_{ir}^b\} \quad v_r \notin M$, otherwise

$$p^k(e_{ij}) = \begin{cases} \dfrac{\tau_{ij}^a \lambda_{ij}^b}{\displaystyle\sum_{r=1}^{d} \tau_{ir}^a \lambda_{ir}^b} & \text{if } v_r \notin M \\ 0 & \text{otherwise} \end{cases} \tag{8.12}$$

The ACS algorithm differs from the ant system by the following: The probability of selecting the shortest edge with the largest pheromone deposit favors exploitation of the search, pheromone concentrations are updated by the ants while constructing solutions using a local updating rule, and at the end of every iteration the global best ant updates the pheromone concentration on the trails using the updating rule.

In the **elitist ant system** the global best ant deposits pheromone on the trail at the end of every iteration in addition to the pheromone deposited by the other ants. In the **rank-based ant system,** a fixed number of best ants are allowed to update the pheromones. The solutions are ranked according to their length. Solutions with shorter lengths (path) deposit more pheromones and solutions with longer lengths deposit less pheromones.

For multi-objective optimization problems, the ACO can be applied by taking a weighted combination of the multiple objectives. If this is not possible, then a set of non-dominated solutions that lie on the Pareto Front can be found by the ACO algorithm. For continuous optimization problems, the search space can be divided into discrete bins and the ACO algorithm applied, or the algorithm can be modified to find the optimum solution for continuous or a combination of continuous-discrete search spaces. Ant colony behavior has been applied in distributed control systems for multiple robots that cooperatively perform a task. ACO has proved itself to be one of the promising algorithms to solve NP-hard combinatorial optimization problems with quality solutions in practical finite time. ACO has been successfully applied to TSP, routing in computer networks, job scheduling, resource management, and recently to machine learning. ACO can be applied to problems where the search space changes dynamically or the variables involved in the problem are stochastic and parallelization is involved. For example, in the TSP the length of edges or the number of nodes could change during the running of the algorithm, and the algorithm should be adaptive to the changes that take place dynamically.

8.5 Summary

ACO has the ability to rapidly converge on the optimum solution for discrete combinatorial problems. ACO is suitable for problems that have an objective function that is to be either maximized or minimized, a set of constraints, and a set of variables or parameters whose values decide the optimum solution. When the set of decision variables is large, an exhaustive search (searching for the optimum solution with all possible combinations of variables) could be intractable or quite impractical. In such cases the ACO algorithm will be suitable for finding the optimum solution in finite time. ACO is more suited to problems that involve finding the shortest path in a graph like network routing since the ants trace a trail through the entire network as a swarm till they find the shortest path.

ACO is suitable for problems that involve finding the shortest path in a network, of which the classical example is the TSP. Other NP-hard problems for which the dimension of the problem (graph size) increases exponentially or larger networks can be solved efficiently with ACO rather than other greedy algorithms. The problems could have a graph or network representation where the characteristics of the graph might change with time. These could be characteristics such as graph connections (edges in TSP), cost associated with each edge, etc. The architecture of the problem (graph) is spatially distributed for which the ACO is suitable.

References

1. Marco Dorigo, Gianni Di Caro, Ant colony optimization: A new meta-heuristic, *Proceedings of the Congress on Evolutionary Computation CEC 1999*, Vol. 2, pp. 1470–1477. IEEE, 1999.
2. Christian Blum, Ant colony optimization: Introduction and recent trends, *Physics of Life Reviews*, Vol. 2, pp. 353–373, 2005.
3. Marco Dorigo, Thomas Stutzle, The ant colony optimization metaheuristic: Algorithms, applications and advances, Chapter 9, In *Handbook of Metaheuristics*, pp. 251–285, April 2006.
4. Marco Dorigo, Mauro Birattari, Thomas Stutzle, Ant colony optimization, *IEEE Computational Intelligence Magazine*, Vol. 1, No. 4, pp. 28–39, November 2006.
5. Marco Dorigo, Christian Blum, Ant colony optimization theory: A survey, *Theoretical Computer Science* (Elsevier), Vol. 344, pp. 243–278, 2005.
6. Marco Dorigo, Thomas Stutzle, *Ant Colony Optimization*, MIT Press, Cambridge, MA, 2004.
7. M. Dorigo, L. M. Gambardella, Ant colonies for the traveling salesman problem, *BioSystems*, Vol. 43, No. 2, pp. 73–81, 1997.
8. M. Dorigo, V. Maniezzo, A. Colorni, Ant system: Optimization by a colony of cooperating agents, *IEEE Transactions on Systems, Man, and Cybernetics—Part B: Cybernetics*, Vol. 26, No. 1, pp. 29–41, 1996.
9. T. Stützle, H. H. Hoos, MAX–MIN ant system, *Future Generation Computer Systems*, Vol. 16, No. 8, pp. 889–914, 2000.
10. L. M. Gambardella, M. Dorigo, Solving symmetric and asymmetric TSPs by ant colonies, *Proceedings of the 1996 IEEE International Conference on Evolutionary Computation (ICEC'96)*, T. Baeck et al. (eds). Piscataway, NJ: IEEE Press, pp. 622–627, 1996.

9

Bee Colony Optimization

9.1 Introduction

The study of bird and insect behavior has led to the development of swarm intelligence techniques for solving optimization problems. The collective intelligence of a swarm of agents (animals, birds, or insects) has proved to be very powerful in solving design problems that have been found to be intractable by classical algorithms. The collective, self-organized group behavior arises from the disciplined behavior of individual members of the swarm in following simple rules, and their study has motivated the development of swarm intelligence algorithms. There is usually a complex social interaction between the members of the swarm that takes care of foraging, sharing, communication, defense against predators, and mating. The members perform activities like foraging for food, building nests, mating, producing, and taking caring of their offspring. Such activities are natural and have been going on for millions of years ever since the population evolved. One interesting characteristic of these swarms is that they are adaptive and flexible and hence survive in the dynamically changing hostile environment. They are robust and together guard against or fight predators, enabling them to survive. The insects follow simple rules as individuals that lead to organized behavior of the entire swarm, and this interaction is important in solving complex problems. They also communicate either by direct or indirect means, and this social structure among the colonies of swarms can accomplish much more than what any single member can achieve. The members not only interact with each other but also with the environment. The swarms usually have a minimum number of individuals in their colonies to take care of all the activities of the entire swarm and exist as a group. Based on these activities there is positive as well as negative feedback that is applied in improvising themselves. There is also a random component in the movement of the swarm members that enables them to explore new areas during their search. Incorporating all this in optimization has led to the development of powerful metaheuristic algorithms that has been proved to solve NP-hard problems with reduced time complexity.

Bees are insects that exhibit collective intelligence and behavior in gathering nectar and producing honey. Study of their intelligent behavior, characteristics, and activities has motivated researchers to develop a set of optimization algorithms to solve complex problems found to be intractable. The original algorithm proposed based on bee intelligence was the bee system in 1997. Since then several variations of the algorithm have been developed and applied to different problems in engineering and computer science. The TSP is one of the famous problems for which a feasible solution has been produced by the swarm intelligence algorithms based on the foraging and mating behavior of honey bees. The bee colony optimization (BCO) algorithm [1, 2] described here is a variant of the bee system with some improvements. The BCO algorithm has been developed based on the foraging behavior of honey bees, and it has both exploitative and explorative abilities to find the global optimum solution to problems [3, 4]. The insects are able to overcome challenging situations and environmental conditions and survive accordingly. This leads to successive generations of swarms carrying out such activities that have been labeled as intelligent. One of the main advantages of such swarm-based algorithms is that the search for the global optimum solution takes place in parallel by multiple members of the swarm, thus reducing search time, and moreover they are able to find good-quality solutions. In the following sections, the foraging behavior of honey bees and the BCO algorithm which is based on these characteristics have been described, followed by variants of the algorithm that depend on other characteristics of honey bees [5].

9.2 Honey Bee Characteristics

Honey bees are insects that can fly, are native to Eurasia, and have also spread to a few other continents. There are 7 to 8 species of honey bees with around 44 subspecies. They form a small fraction of the tens of thousands of bee species (~20,000), and they belong to the genus *Apis*. Honey bees are famous for collection and storage of honey in hives that have been the targets of birds, animals, and human foragers. They build nests (hives) from wax, produce and store surplus honey, and live in large colonies. They are largely exploited for their honey and wax. They are good pollinators of flowering species of plants since they sit on flowers and collect nectar. Figure 9.1 shows a honey bee that is completely covered in pollen from sitting on a dandelion flower, and in the process the bee carries pollen on its body.

Figure 9.2 shows a cryptic bumblebee (*Bombus cryptarum*) sitting on the European goldenrod flower in Northwestern Estonia and in the process carrying out pollination. Figure

FIGURE 9.1
Pollination of dandelion flower by bee. (Author: Guerin Nicolas – own work, CC BY-SA 3.0. https://commons.wikimedia.org/wiki/Commons:GNU_Free_Documentation_License,_version_1.2. https://creativecommons.org/licenses/by-sa/3.0/deed.en.)

FIGURE 9.2
Bumblebee on the European goldenrod. (Author: Ivar Leidus – own work, CC BY-SA 4.0. https://creativecomm ons.org/licenses/by-sa/4.0/deed.en.)

9.3 shows an Italian bee contributing to the pollination of the white sweet clover flower that contains yellow pollen in Northwestern Estonia.

One colony of bees consists of a queen bee, several male drone bees, and a large number of female worker bees. The colony size could vary from hundreds to thousands of honey bees. Reproduction is by laying of eggs in the cells of the wax honeycomb. Figure 9.4 shows the honeycomb of the honey bees made of wax containing eggs and larvae of drone bees that are around three to four days old. Figure 9.5 shows a nest of honey bees consisting of hundreds of honey bees on the branch of a tree.

The queen and worker bees are developed from fertilized eggs whereas the male drone bees are developed from unfertilized eggs. Larvae are fed with honey and pollen, and the one fed with royal jelly develops into the queen bee. The queen bee is the biggest bee in the colony, and its lifespan is a few years, normally less than ten. The other bees live for several months, typically less than one year. The drone bees die after mating with the queen. Worker bees are responsible for cleaning the hive, producing wax cells, feeding the larvae, guarding the hive, and receiving pollen and nectar from the foragers. The forager worker bees perform a waggle dance to communicate to other members of the hive about having found food, its location, and quantity. Figure 9.6a shows some of the bees performing a waggle dance on the dance floor of the hive. A waggle run oriented 45° to the right of 'up' on the vertical comb indicates a food source 45° to the right of the direction of the sun outside the hive. Figure 9.6b illustrates the orientation of the food source with respect to the position of the sun and the direction of the bee dance.

In the process of collecting nectar from flowers, bees also aid in pollination. The wax used in building the bee hive is also targeted by humans for making several crafts. The hives are normally built in trees or plant shrubs and sometimes in buildings. A hive usually consists of one female bee called the queen bee and several hundreds of drone bees and worker bees.

FIGURE 9.3
Italian bee pollinating white sweet clover. (Author: Ivar Leidus – own work, CC BY-SA 4.0. https://creativecomm
ons.org/licenses/by-sa/4.0/deed.en.)

FIGURE 9.4
Honeycomb of honey bees. [Author: Waugsberg (talk – contribs), CC BY-SA 3.0. https://commons.wikimedia
.org/wiki/Commons:GNU_Free_Documentation_License,_version_1.2   https://creativecommons.org/license
s/by-sa/3.0/deed.en.]

FIGURE 9.5

Nuclei of honey bees' nest on a branch. [Author: Stolz Gary M, U.S. Fish and Wildlife Service (Public Domain).]

The worker bees become scouts or foragers when they leave the hive to look for food sources. The worker bees scout for food, and once it is found, they fill their stomach with nectar and return to the hive to deposit the nectar in the wax honeycomb along with an enzyme secreted to produce honey. The discovery of a food source is communicated to other bees by means of a waggle dance. The waggle dance is a special form of dance performed by the bees on the dance floor of the hive. The duration and direction of the dance indicates the quality, direction, and distance of the food source from the hive. Bees also perform round and tremble dances as different forms of communication. Honey bees perform tremble dance to communicate to the worker bees to collect nectar from forager bees. Bees have good navigation capabilities

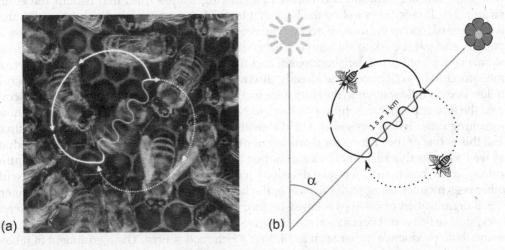

FIGURE 9.6

(a) Figure-of-eight-shaped waggle dance of the honeybee. [Author: (Figure design: J. Tautz and M. Kleinhenz, Beegroup Würzburg). Source: Chittka L: *Dances as Windows into Insect Perception*. PLoS Biol 2/7/2004: e216. https://dx.doi.org/10.1371/journal.pbio.0020216. https://creativecommons.org/licenses/by/2.5/deed.en.]. (b) Bee dance. (Source: file: bee.dance.png. File: sun01.svg. File: abeille-bee.svg by Emmanuel Boutet. File: RosendeutschschweizerBlatt.svg by Kilom691. https://creativecommons.org/licenses/by-sa/2.5/deed.en.)

and memory so that they can fly out from the hive several kilometers while looking for flower patches (food sources) and then come back to the hive remembering the location of the food source as well as the direction and distance. This requires some mapping of the environment in space within the memory of the bees.

Queen bees fly out from the hive and mate with drones outside their colony. Before establishing a colony, the bees scout for a good location and then the queen and the workers establish their hive after collectively deciding on the location. They build their honeycomb from wax and brood worker bees. Bees mostly survive on pollen and nectar they collect from flowers. Honey bees are the only species among bees to have small barbs on the sting. When there are intruders in the colony, bees sting them and communicate to other bees (raise the alarm) by secreting a chemical. Their communication is by means of secreting chemicals and performing dances. The orientation of the bee while dancing indicates the direction of the food source with respect to the sun position. Honey bees cannot withstand cold temperatures, especially less than 10°C. At those temperatures, they crowd in their hive. In some places, beekeepers transport bee hives (along with bees) to take care of pollination requirements of agricultural fields. The bees collect nectar (ingest), process it, and put it in honeycombs to become honey. Worker bees secrete wax from their abdomen glands that forms the walls and caps of the comb. The scout bees are employed for foraging, and they can fly a few kilometers to forage for food, looking for nectar in flower patches. Even after discovering a food source, they scout for better quality flower patches. They go back to the hive and communicate the information of the find to the other bees by performing a waggle dance.

As shown in Figure 9.6a, the waggle dance of bees has an approximate shape of the digit 8. There is usually a dance floor in the bee hives. The scout bee moves in a straight line in a direction that is relative to the sun's azimuth to indicate the direction of the food source with respect to the location of the bee hive, as shown in Figure 9.6b. Then it moves in an alternating left and right return path in an almost circular trace. The speed and duration of the dance indicate the distance to the food source, and the frequency of the waggles is an indication of the quality of the food source. There is a large variation in behavior of different species but all of them exhibit collective intelligence and accomplish complex tasks with their self-organization and nature of following simple rules that benefit the entire swarm. The forager bees follow the bees that have already discovered a flower patch that is a source of food in the form of nectar. The bees procure nectar from the discovered flower patches and get it back to their hive for the storage and production of honey. If the food source is depleted it will be abandoned, and the foragers look for new sources. If there is more food to be collected at the already discovered site, the bees could dance and recruit fellow bees to follow them to the source for further collection of nectar. The bee that discovered the flower patch can continue its foraging and collection of nectar at the patch without recruiting other fellow bees also. The dance also indicates the quality of the food source, and this is one of the criteria for recruitment of fellow bees. This demonstrates individual as well as collective social behavior of honey bees that is for the betterment of the entire colony in the bee hive. They take decisions individually as well as in collaboration with other bees by interacting with the bees in the hive in a very adaptive and flexible manner.

Self-organization of colonies is based on four principles: Positive feedback, negative feedback, stochasticity, and population size. Positive feedback enhances an activity that has been found to be productive earlier such as finding a rich food source. The recruitment of fellow foragers to such a site is an example of positive feedback. This is done by laying pheromones in ant colonies and by dancing and recruiting fellow foragers in bee colonies. Negative feedback is necessary for maintaining balance in the colony and in the environment. If a food source is not plentiful or of good quality for consumption, the ants do not lay pheromone on the way back. Similarly, the bees do not dance and do not recruit fellow bees to the site or the site is abandoned by the scout bees. Randomness in the areas of search or uncertainty is essential for the exploration of new and uncharted spaces leading to better food finds or solutions. If the fellow insects search in the same area that has been covered earlier, it might not be productive since the food source might have depleted because of consumption. There will be a limited number of food sources within a neighborhood or in a small local area, and repeated search in this region might not be fruitful. Expanding the search to new unexplored territories could be beneficial both in terms of quality as well as quantity of food sources that are not only plentiful but also rich. This involves some uncertainty in the search since diversifying the search in these uncharted areas could also turn out to be unproductive. For all these searches, a population of insects or agents is required that conduct the search in parallel. Unless this inherent parallelism by a population of members is included in the search, it will take a long time to cover the entire region in the search for food. The population survives attacks by predators and changing environmental conditions that could also be hostile because of this group dynamism and self-organization. There is usually a hierarchy within the population that is responsible for decentralization and control. All the members follow the rules, benefiting the entire swarm. The swarm-based optimization algorithms are population-based, and most of them undergo multiple iterations. Every member of the population ends up with a solution to the problem in each iteration. At the end of all the iterations, there are N number of solutions to the problem, where N is the population size. The best of these solutions is the global optimum or an approximation of it. This is attained by the value of the objective or quality function that is mathematically modeled based on the application.

9.3 Bee Colony Optimization

The bee colony optimization (BCO) algorithm has been developed from the social interactions and natural foraging behavior of honey bees. The algorithm makes use of both the exploration and exploitation properties of honey bees. The collective intelligence of the bee swarm is capable of solving complex combinatorial optimization problems. The bees are very methodical in collecting and storing nectar and processing it to produce honey. The scout bees go in search of flower patches (food sources). If they discover a food source they make an assessment of the quality and quantity of the food source at the flower patch discovered. They take a load of nectar from the flowers and go back to the hive. Once they reach the hive, the collected load of nectar is stored and then the honey bee performs a waggle dance on the dance floor. The direction of and duration of the bee dance is the dissemination of information to the other bees about the discovery. The scout bee passes on information about the distance of the food source from the bee hive, and its quantity and quality. This could make fellow bees follow the scout bee to the food source (recruitment) or the scout bee could go back alone to the discovered food source if it is found to be abundant and rich. If the source is not plentiful, the scout bee could decide to abandon it and forage elsewhere.

9.3.1 Algorithm

The BCO algorithm starts with an initial group of scout bees (foragers) of population size N. The bee population is assumed to be present in the hive initially. The feasible solutions are constructed by the bees, incrementally adding solution components, in a manner that is similar to the ACO algorithm. The bees start with partial solutions and build up a complete solution to the problem in an incremental manner. The BCO algorithm is iterative, and the maximum number of iterations can be chosen depending on the application. When one or more bees have a complete feasible solution, one iteration ends. Also, at the end of every iteration there are not only one or more complete feasible solutions but also partial solutions constructed by other bees. Moreover, at the end of every iteration, the best among the complete solutions built by the bees is saved in memory. In a succeeding iteration, if any solution is found to be better than the current best, it replaces the current best solution in memory. At the end of the maximum number of iterations, the best solution among all the solutions attained so far is the global optimum. If there is a stopping criterion defined for the problem, when the algorithm satisfies the stopping criterion the current best solution is the global optimum.

There are two passes for every bee in each iteration, the forward and the backward pass. The bees leaving the hive to either known (already discovered) or unknown food sources is the forward pass. The bees returning to the hive either with nectar or without nectar is the backward pass. This is equivalent to the bees assembling solution components to create partial or complete solutions during the forward pass. During the forward pass the bees use their individual intelligence, the past history of already discovered food sources, and the quality and quantity of the food source already discovered in order to construct one or more partial solutions. During the backward pass, the collective intelligence and communication (information exchange) of the bees lead to decision-making in the hive. When bees are starting to forage or during any iteration when they abandon a food source, they might use local exploration to discover new food sources. Following a forward pass there is a backward pass for every bee. During the backward pass the bees go back to the hive and exchange information about the quality of the food sources discovered. This is the time when the bees perform the waggle dance to communicate to the other bees about the quality and quantity of the food source discovered. The bees in the hive participate in a decision-making process based on the information exchanged. They decide whether to recruit bees to continue foraging or abandon the already discovered site, or forage there individually without recruiting other bees. This is equivalent to exchange of information about the partial solutions created by all the bees. The decision whether to continue with the already assembled partial solution or ignore it and start creating a new partial solution or follow the partial solution created by another bee is taken at the end of the backward pass by all the bees in the hive. This decision is taken based on the quality of the partial solutions created. This is followed by another forward pass where the already created partial solution is expanded and then a backward pass where the bees return to the hive to repeat the evaluation of the partial solutions and engage in decision making. This alternate forward and backward pass mechanism of the bees continues until one or more complete feasible solutions are created. When there is at least one complete feasible solution to the problem, one iteration ends. Thus every iteration consists of multiple forward and backward passes undertaken by the bees.

Figure 9.7 illustrates the forward pass undertaken by the artificial bees. There is a bee hive and six bees are shown, and they are either inside or coming out of the hive to search for good quality food sources which is equivalent to assembling good feasible solutions to the problem incrementally. These incremental solutions are built by the bees in multiple stages during multiple forward passes. Every stage has multiple partial solutions built by the various bees. The total number of stages is assumed as K. There are several nodes in each stage. It is assumed that there are n number of nodes in each stage. During each forward pass the various scout bees go to the succeeding stages to incrementally build the partial solution, i.e. they advance stage by stage in each forward pass of one iteration.

Figure 9.8 illustrates the backward pass of the artificial bees going back to the hive. The backward pass follows the forward pass of every iteration. Thus there are multiple pairs of forward and backward passes during every iteration until the maximum number of iterations is

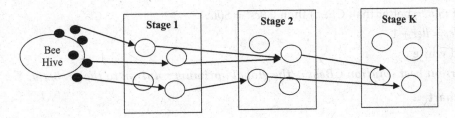

FIGURE 9.7
Forward pass of the artificial bees.

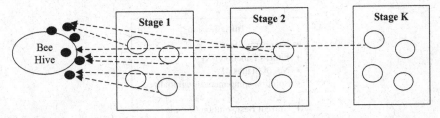

FIGURE 9.8
Backward pass of the artificial bees.

reached. During the backward pass the bees go back to the hive from whichever stage they are in. Here it is assumed that the number of forward/backward passes is equal to the number of stages K. The incremental solutions are built in every stage by foraging (searching) at the nodes of that stage. Consecutively, the bees could continue foraging on their individual paths chosen or they could join another bee in its path if the quality of the food source is good enough. Such decisions are taken by the bees when they return to the hive during the backward passes. Some of the bees could recruit fellow foragers before they continue to build their partial solutions whereas others will continue alone. Some of them might abandon their partial solutions already built and start building new solutions in the next forward pass of the current iteration.

9.3.2 Pseudocode

Initialization

Population size of bees N

Define objective function for the problem $f(X)$

Number of stages M

$f(X)$ consists of solution components $[X_1, X_2, ..., X_M]$

Current best solution $CBest$ is initialized to any feasible solution

Maximum number of iterations $MaxIter$

$n_p = 1$ (n_p is variable for indexing the stages)

$iter = 1$

while *(iter $\leq$ MaxIter)*

for $n_p = 1$ to M

Bees fly from the hive and create partial solutions during forward pass

The partial solutions are chosen from the set available at stage $s(n_p)$ Bees go back to the hive during the backward pass and information is exchanged

end for

if (*S*(*iter*) better than *CBest*) **then** *CBest* = *S*(*iter*)

    iter = *iter* + 1

end while

Current best solution CBest is the global optimum solution to the problem.

Flowchart

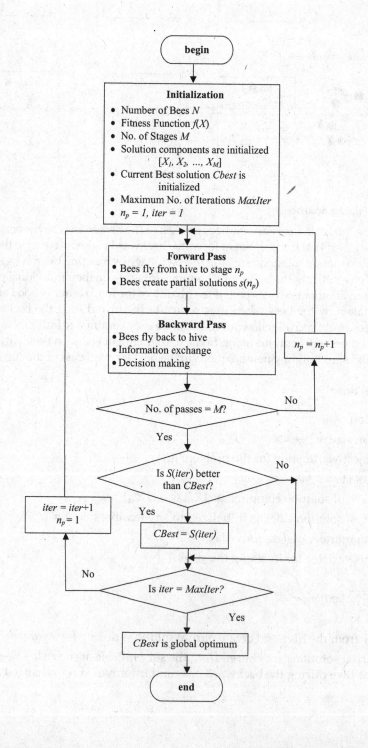

9.4 Variants of the Algorithm

The variants of the BCO algorithm are developed based on the characteristics, social interaction, and complex behavior of honey bees. The foraging behavior of honey bees has led to the development of the bee system (BS), bee colony optimization (BCO), artificial bee colony optimization (ABC), and the bees algorithm (BA). The mating and breeding behavior of honey bees has motivated the development of marriage in honey bees optimization (MBO), fast marriage in honey bees optimization (FMHBO), and honey bees mating optimization (HBMO). The bee hive algorithm has been developed from the study of the behavior and communication of the bees within the bee hive. Some variants of these algorithms are proposed from the way in which the decision-making process of the bees is carried out. The decisions include visiting the nodes, following the same path or not in the next forward pass, recruiting fellow bees to follow them, abandoning a site already visited, and so on. One of the notable variants of the algorithm is the fuzzy bee system (FBS) and the fuzzy bee colony (FBCO) algorithm that incorporates fuzzy logic and approximation in reasoning when there is uncertainty in the problem.

The **bee system (BS)** was proposed by Sato and Bagiwara (1997) and is an improved and extended version of the genetic algorithm (GA). The bee system is based on the foraging behavior of honey bees [6]. When a bee finds good-quality food, it recruits other bees by performing a dance and returns to the discovered food site and searches around the neighborhood looking for even better food sources. The bee system uses a combination of local as well as global search and applies what is known as a concentrated crossover operator that is an extension of the crossover operator in GA. Concentrated crossover intensifies the search around the areas where the global optimum could possibly exist. The simplex algorithm has been modified to be termed as pseudo-simplex, and has been applied here to enhance the local search abilities of the algorithm. The two main advantages of GA in having a derivative-free global search with less possibility of getting trapped in local optima have been utilized in developing the bee system.

GA uses a population of agents to search for the optimum solution whereas the bee system uses multiple sets of populations for global and local search. Initially the global search is undertaken by the global search population and the chromosomes with high fitness values are chosen. This high fitness value chromosome is equivalent to a bee that has discovered a good-quality food source during foraging. The number of such high fitness value chromosomes is C_{max}. The local search is done intensively around these chromosomes that were selected during the global search. For this local search multiple local populations are employed where the number of such populations is equal to C_{max}. In concentrated crossover, all the chromosomes of the local search population assigned to chromosome C_r undergo crossover with the selected chromosome C_r. Therefore higher fitness values are propagated by means of concentrated crossover to all the members of the local search population. Another operation introduced in this bee system is migration wherein one member of every local population is transferred to a neighboring population. This is followed by the pseudo-simplex operation which uses only three points in the search space irrespective of the dimension of the space. This is a simplified version of the simplex method. The local search ends after completing the fixed number of generations. If the stopping criterion or the global optimum is attained, the algorithm stops. Otherwise the next iteration starts and continues until the maximum number of iterations is completed. Compared to the conventional GA, the bee system has been found to give better performance.

The **artificial bee colony (ABC) algorithm** has been developed from the intelligent behavior of honey bees [7]. The natural bees are the inspiration behind the artificial bees used in the algorithm. A population of artificial bees is generated and they are classified into three categories: *employed, onlookers, scouts*. An *employed* bee visits a food source discovered by itself during an earlier search. An *onlooker* bee is one that is waiting by the side of the dance floor to make a decision on whether to visit a food source or not. A *scout* bee is one who does a random search looking for unexplored new food sources. The artificial bees are classified into *employed* and *onlookers* initially. Every discovered food source is associated with one *employed* bee. When the food source is depleted the *employed* bee turns into a *scout* bee. The algorithm is iterative, with the number of *employed* and *onlooker* bees being almost equal (approximately half the bee population) and one *scout* bee per iteration. The quality of the food source is evaluated by its richness of nectar, and this is the fitness or objective function of the problem. The probability of an onlooker bee selecting a food source is directly proportional to its richness. The locations of food sources are the possible candidate solutions to the problem that are initially selected randomly. Therefore the number of *employed* bees is equal to the number of food sources chosen and this is again equal to the number of *onlooker* bees. During each iteration, the *employed* bees convey the information about the discovered food source to the *onlooker* bees by performing a waggle dance, and the *employed* bees return to the original food site and look for new food sources in the neighborhood based on visual inputs. If the new neighboring food source is richer than the previous one, the old one is replaced by the new one. The depleted and abandoned food sources are replaced by new ones by the *scout* bees. Thus the *scout* bees take care of exploration of the search space whereas the *employed* and *onlooker* bees exploit the local neighborhood of the existing solutions. The speed of discovery of new and better solutions leads to a faster rate of convergence of the algorithm.

The **bees algorithm** has been proposed based on the study of the foraging behavior of honey bees [8, 9, 10]. The algorithm combines exploitation of the local neighborhood as well as exploration of the search space to find the global optimum. The algorithm starts with an initial population of N bees. It is assumed that there are flower patches (food sources) in the search space, and the total number of such flower patches is assumed as F_P. An objective (fitness) function $f(X_i)$, $i = 1, 2, ..., F_P$ is defined whose value is proportional to the quality of the flower patch on which it is evaluated. The fitness function is evaluated on every flower patch that has been initially selected in the search space. The evaluated fitness function values are ranked in descending order. From these ranked fitness values, M number of patches with the highest fitness values among the F_P are selected. The selected M flower patches are divided into elite and non-elite sites. Let M_E be the number of elite flower patches and M_{NE} be the number of non-elite flower patches. The algorithm applies a local search in the neighborhood of the flower patches chosen based on their ranked fitness values. For this neighborhood search, forager bees are recruited both for the elite as well as non-elite sites. Let the neighborhood size be S_N, the number of forager bees recruited for elite sites be N_E, and the number of forager bees recruited for non-elite sites be N_{NE}. The number of patches which are lower in the ranking, that is, that are not in the top M ranks chosen, is M_L ($F_P = M + M_L = M_E + M_{NE} + M_L$). The local search is in the neighborhood of the elite and non-elite sites chosen whereas the global search includes the M_L sites that were not selected because they were lower in the ranking. The number of forager bees recruited for the M_L sites that were lower in fitness ranking is ($N - N_E - N_{NE}$). As a last step, overall sorting of the fitness values is done and the iterations are repeated until the global optimum is attained. A threshold function could also be defined that could serve as the stopping criterion for the algorithm instead of running the maximum number of iterations.

The **marriage in honey bees optimization (MBO) algorithm** is based on the mating and breeding behavior of honey bees [11]. Initially there is a single queen bee and later on there is a colony of queen and other bees (drones, workers) that models the marriage behavior in honey bees. There is usually only one queen bee in a hive, but some species might have more than one queen in their hive. The queen bee is responsible for laying eggs, either fertilized or unfertilized, and brooding. The drones are the male bees that mate with the queen to produce eggs. The mating takes place during flight, away from the hive. One queen can mate with up to 20 drones at a time. The queen bee performs a dance to indicate the start of mating flight. The worker bees take care of the hive and the brood and sometimes they lay eggs. The unfertilized eggs become the drones, and the fertilized eggs develop into queens and workers. During mating the sperms are put in a genetic pool called spermatheca, and each time the queen lays an egg it is fertilized from the sperms in this genetic pool. Successful mating (adding of a sperm to the spermatheca) depends probabilistically on the fitness of the queen and drone and the speed of flight of the queen bee. The mathematical equation modeling this probability is given by Equation 9.1.

$$p(Q, D_r) = e^{-\frac{\Delta F}{QueenSpeed(n)}} \tag{9.1}$$

where $\Delta f = |f(D_r) - f(Q)|$, and $f(D_r)$ and $f(Q)$ are the fitness of the *rth* drone bee and the queen bee respectively. Initially when the queen bee starts on her flight her energy and speed are high and the successful mating probability is high. This is gradually reduced during the flight modeled by the following equations:

$$QueenSpeed(n+1) = rand \cdot QueenSpeed(n) \tag{9.2}$$

where rand is a random number that can assume values in the interval [0, 1].

$$QueenEnergy(n+1) = QueenEnergy(n) - EnergyStep \tag{9.3}$$

where *EnergyStep* is the step size with which the energy of the queen is reduced with each transition. It is assumed that the queen bee takes flight and changes state in what is called a transition and the energy is reduced by the quantity *EnergyStep* with every change of state. This mating continues until the energy of the queen reduces to a minimum level or the spermatheca is full. The speed, energy, and position of the queen are initialized in space with random values. The steps described above for mating take place for all the queens if there is more than one queen in the bee hive.

When the queen bee returns to the hive it starts breeding by crossover operation of its genome with any one of the randomly chosen sperms from its spermatheca. The crossover operations generate a brood. The mutation operator is applied on these broods to introduce diversity into the population. Mutation also prevents generation of the same brood in case the same sperm is chosen twice from the spermatheca for crossover. The worker bees improve the fitness of the brood thus produced. The fitness of the worker bee is increased in proportion to the improvement in the fitness of the brood. Thus, after mating, the queen bee generates a set of partial solutions. The worker bees are allotted to take care of the brood generated by crossover and mutation. The queen with the least fitness function value is replaced by the brood with the highest fitness value and this is repeated until there is no queen which is less fit than any of the broods. The remaining broods are eliminated, and the next mating flight takes place.

The workers and queens are initially chosen randomly with the number of queens being much smaller than the number of workers. The energy, speed, and position of the queens are initialized. The number of mating flights (iterations) is initialized. The queens make state transitions and choose the drones to mate with and fill their spermatheca according to Equation 9.1. The energy and speed of the queen are modified according to Equations 9.2 and 9.3. When one mating flight is over, broods are generated by applying crossover of sperms with genomes, followed by mutation. The workers are applied to improve the fitness values of the broods. The queens with the least fitness values are replaced by broods which have a better fitness value. The remaining members of broods are deleted. The next mating flight takes place with the new queens in the new iteration, and this continues until the maximum number of iterations is completed. The number of queens can go from 1 to 5, the number of matings per flight of 1 queen can range from 5 to 20 (spermatheca size), and the number of broods generated by 1 queen can range from 20 to 100, the number being inversely proportional to the number of mating flights. These are some of the typical values used in experiments.

The **fast marriage in honey bees optimization (FMHBO)** algorithm is an improvement over the MBO algorithm in being faster in convergence for global optimization [12]. In MBO, there are three mathematical equations for the computation of mating probability, speed, and energy of the queen bee, and this takes up time during each iteration. In FMHBO, these computations are avoided by randomly choosing a drone to mate with the queens instead of calculating the probability during each iteration. Moreover, the energy and speed calculations of the queen are not done in order to save computation time, making the algorithm faster than the previous version of MBO. There are three operators in FMHBO: *crossover*, *mutation*, and *heuristics*. *Crossover* and *mutation* are already used in MBO, whereas the *heuristic* operator is applied here to conduct a local search on the broods and improve their fitness.

The **honey bees mating optimization (HBMO)** algorithm was inspired by the mating behavior of honey bees [13]. The queen bee mates with the drone bees during the mating flight. The drone bees are large in number (hundreds), and the mating takes place tens of times. The drones are haploids, i.e. they contain only the mother's gamete and this is propagated without any alteration. Chromosomes in egg or sperm cells are called gametes. The queen lays eggs and the sperms are retrieved from the spermatheca to fertilize the eggs. The queen starts the mating flight with a dance which is a communication to the drones to follow her for mating. The mating of the queen bee with the drones takes place in mid-air, and every drone dies after mating. This type of mating is a unique characteristic of honey bees. The probability of adding the sperm of the drone D_r ($r = 1, 2, ..., P$) to the spermatheca is given by Equation 9.1, where P is the population size of the drones that are going to mate with the queen Q. The probability of productively mating (adding sperm to spermatheca) is high initially during the mating flight since the fitness of the drone (proportional to energy) almost matches that of the queen and the speed of the queen is also high. After each mating with a drone the energy and speed of the queen reduce as given by Equations 9.2 and 9.3.

The HBMO algorithm is similar to the marriage in honey bees optimization algorithm (MBO) in that both the algorithms are developed from the mating behavior of honey bees. The mating flight of the queen bee starts with the typical dance indicative of the start of mating flight. The probability of successful mating with a drone is given by Equation 9.1. After each successful mating of the queen with a drone from the set of drones (population size P) the sperm is added to the spermatheca of the queen whose capacity is limited (maximum number of sperms it can hold is fixed initially). The spermatheca size is usually chosen as the number of matings per flight of the queen. Broods are created by a crossover operation between the eggs laid by the queen and the sperms introduced by the drones (selected randomly from the sperms stored in the spermatheca). These are the partial solutions to the problem. The worker bees conduct a local search among the broods and could be made to improve upon the fitness of these broods by applying heuristics. The fitness function is evaluated on these broods and the queen, and if a brood is found to be fitter than the queen, then the queen is replaced by the corresponding brood whose fitness is higher than that of the queen. This process is repeated for the queens and one iteration ends. The remaining broods (ones that do not replace the queens) are eliminated. In the next iteration, again the above process of mating is repeated until the maximum number of iterations is reached or a stopping criterion is met. In [14] the honey bees mating algorithm is applied to solve the non-linear Diophantine equation benchmark problem and its performance is compared with GA. In addition, the algorithm has been applied to guide a mobile robot through space with different obstacles.

The **fuzzy bee colony optimization (FBCO)** algorithm is a variant of the BCO algorithm that incorporates fuzzy logic in the foraging behavior of honey bees [15]. The parameters of the BCO algorithm are tuned by applying fuzzy logic to improve the performance. In the **fuzzy bee system (FBS)** the bees use fuzzy reasoning and approximate rules in communication and other decision-making activities. The fuzzy membership functions are used in quantifying the uncertainties associated with the problem. The advantage of combining fuzzy logic with honey bee behavior-based optimization algorithms is that it is capable of solving both deterministic problems (where all the parameters and constraints of the problem are known) as well as problems with which there is some uncertainty associated. In [16] novelty is introduced in clustering by incorporating fuzzy rules in the honey bee foraging optimization algorithm. Clustering is one of the important techniques in data mining, pattern recognition, and image classification. Several hundreds of published works are available for the BCO and its variants with applications related to continuous and discrete optimization problems. This chapter has covered only a fraction of those applications for which the BCO has been applied.

9.5 Summary

The BCO algorithm is capable of solving difficult combinatorial optimization problems. It is a metaheuristic algorithm that has been inspired by the behavior of honey bees in nature. The performance of the algorithm shows that simple insects like bees can inspire and motivate us to develop algorithms to solve complex problems. Rather than their individual intelligence, the collective intelligence exhibited by the swarm as a whole and the dynamic and adaptive nature of the bees is responsible for such outstanding performance. Their individual performance and interactions amongst each other are for the benefit of the swarm as a whole, and hence they benefit every member. The duties of insect colonies are usually distributed among the swarm members in a very optimal way, and they function in a responsible manner. The optimization algorithm based on the behavior of honey bees has been successfully applied to several real-life problems such as job shop scheduling, clustering, image analysis, optimal design of structures, and complex engineering problems. The individual and collective complex behaviors of honey bees have been mimicked in developing these algorithms to solve NP-hard problems.

References

1. D. Teodorović, Bee colony optimization (BCO). In: *Innovations in Swarm Intelligence. Studies in Computational Intelligence*, Vol. 248, C. P. Lim, L. C. Jain, S. Dehuri (eds). Berlin, Heidelberg: Springer, pp. 39–60, 2009.
2. Dusan Teodorovic, Panta Lucic, Goran Markovic, Mauro Dell' Orco, Bee colony optimization: Principles and applications, *IEEE 8th Seminar on Neural Network Applications in Electrical Engineering, NEUREL 2006*, University of Belgrade, Serbia, 25–27 September 2006.
3. Tatjana Davidovic, Dusan Teodorovic, Milica Selmic, Bee colony optimization part I: The algorithm overview, *Yugoslav Journal of Operations Research*, Vol. 25, No. 1, pp. 33–56, 2015.
4. Dusan Teodorovic, Milica Selmic, Tatjana Davidovic, Bee colony optimization part II: The application survey, *Yugoslav Journal of Operations Research*, Vol. 25, No. 2, pp. 185–219, 2015.
5. Dervis Karaboga, Bahriye Akay, A survey: Algorithms simulating bee swarm intelligence, *Artificial Intelligence Review*, Vol. 31, No. 1–4, pp. 61–85, 2009.
6. T. Sato, M. Hagiwara, Bee system: Finding solution by a concentrated search, *Proceedings of the 1997 IEEE International Conference on Systems, Man and Cybernetics*, Orlando, FL, United States, pp. 3954–3959, 12–15 October 1997.
7. Dervis Karaboga, Bahriye Basturk, A powerful and efficient algorithm for numerical function optimization: Artificial bee colony (ABC) algorithm, *Journal of Global Optimization*, Vol. 39, No. 3, pp. 459–471, November 2007.
8. Baris Yuce, Michael S. Packianather, Ernesto Mastrocinque, Duc Truong Pham, Alfredo Lambiase, Honey bees inspired optimization method: The bees algorithm, *Insects*, Vol. 4, No. 4, pp. 646–662, 2013.
9. Ebubekir Coc, The bees algorithm: Theory, improvements and applications, Ph.D. thesis. Manufacturing Engineering Centre, School of Engineering, University of Wales, UK, March 2010.
10. D. T. Pham, A. Ghanbarzadeh, E. Koc, S. Otri, S. Rahim, M. Zaidi, *The Bees Algorithm, Technical Note*, Cardiff, UK: Manufacturing Engineering Center, Cardiff University, 2005.
11. H. A. Abbass, MBO: Marriage in honey bees optimization – A haplometrosis polygynous swarming approach, *Proceedings of the 2001 Congress on Evolutionary Computation (IEEE)*, Seoul, South Korea, 27–30 May 2001.
12. C. Yang, J. Chen, X. Tu, Algorithm of fast marriage in honey bees optimization and convergence analysis, *Proceedings of the IEEE International Conference on Automation and Logistics, ICAL 2007*, Jinan, China, pp. 1794–1799, 18–21 August 2007.
13. A. Afshar, O. Bozorg Haddad, M. A. Marino, B. J. Adams, Honey-bee mating optimization (HBMO) algorithm for optimal reservoir operation, *Journal of the Franklin Institute*, Vol. 344, No. 5, pp. 452–462, 2007.
14. Petar Curkovic, Bojan Jerbic, Honey-bees optimization algorithm applied to path planning problem, *International Journal of Simulation Modeling*, Vol. 6, No. 3, pp. 154–164, 2007.
15. Amador-Angulo L, Castillo O, A fuzzy bee colony optimization algorithm using an interval type-2 fuzzy logic system for trajectory control of a mobile robot, In: *Advances in Artificial Intelligence and Soft Computing MICAI 2015*, G. Sidorov, S. Galicia-Haro (eds), Lecture Notes in Computer Science, Vol. 9413. Cham: Springer, 2015.
16. Ali-Asghar Gholami, Ramin Ayanzadeh, Elaheh Raisi, Fuzzy honey bees foraging optimization: Swarm intelligence approach for clustering, *Journal of Artificial Intelligence*, Vol. 7, pp. 13–23, 2014.

10

Fish School Search Algorithm

10.1 Introduction

Swarm behavior is collective and coordinated activity by a group of similar entities, normally exhibited by animals, birds, and insects. When the animals, birds, or insects stay together in a neighborhood, move together while foraging for food or migrating, and interact with each other in exchanging information, it is swarm intelligence and behavior. Swarm intelligence is the collective intelligence exhibited by a group of animals, birds, or insects that has a social organization and hierarchy. They exhibit individual as well as group dynamics that are for the benefit of the entire swarm as a whole. The members of the swarm interact with each other and with the environment in their everyday activities. The population-based search algorithms are based on these behaviors of swarms. The search for the optimum is undertaken in parallel by several agents in the multidimensional search space. Metaheuristics is another component that makes these algorithms efficient because heuristics simplifies the search to some extent.

Inherent parallelism is one of the main advantages of these nature-inspired optimization algorithms. If the search for the optimum is made exhaustively, it is not very effective and the time complexity of the algorithm increases. When the search space has higher dimensions, the search becomes complex and time-consuming.

Biological systems have been evolving over millions of years, and each species has its own methods of dealing with the complexities of nature. The members of a species follow simple rules individually and also as a group, with social interactions happening among the members of the group. Most of the swarms incorporate delegation of duties, communication among members, and memory to store past successes. The fish school search (FSS) algorithm has been developed from the schooling behavior of fish, an aquatic animal. Fish swarms have hundreds to thousands of fish, and they are suitable for solving problems with unstructured high-dimensional search spaces. They exhibit group dynamism and intelligence while foraging and migrating. These simple creatures have devised their own methods of dealing with the complexities of nature as well as the hostile aquatic environment. They are adaptable, flexible, and they communicate within themselves for the greater good of the swarm. The characteristics and behavior of fish have been adopted into the FSS algorithm that has been described in the following sections.

10.2 Fish School Behavior

Fish are vertebrate aquatic animals that can be found in almost all waterbodies, from flowing streams to lakes and deep oceans. More than 30,000 diversified species of fish have been identified, and they are available in plenty in all types of waterbodies. They use underwater acoustics to communicate with members of their own species. Fish are one of the main sources of food for the human race while some varieties are used for ornamental purposes. Fish are supposed to be 'cold-blooded' because their body temperature changes with the ambient temperature of water. Fish have a streamlined body for swimming that is covered with scales, with multiple pairs of fins and gills for breathing. The size of fish varies from the tiny fish less than 10 mm in length to the big sharks that are 16 to 20 m in length. Fish have jaws for eating, and they feed on plants and other organisms, and the bigger fish sometimes feed on the smaller fish. Fish reproduce by laying eggs that are either hatched outside or nourished inside.

By definition, a group of fish that stay and swim together is *shoaling*, and if they all swim together in the same direction in a coordinated manner it is *schooling*. Figure 10.1 shows *shoaling* of *blue-and-gold snappers* in the Coiba National Park, Panama; they are swimming together connected as a social group but also independently. Figure 10.2 shows *schooling* of the *big eye scad* fish in Hawaii; they congregate in large schools in shallow water during daytime for protection from predators. The school normally has hundreds to thousands of fish [1].

In typical swarming behavior exhibited by fish every individual follows simple rules and also the social norms of the entire group. This ensures an organization or discipline when moving together, enabling the exchange of information related to location of food or prey, guarding themselves against predators, mating, and migration. The fish swarm has many eyes scanning for food and predators. Fish swim in large schools for protection from predators, and one such is shown in Figure 10.3.

FIGURE 10.1
Shoal of *blue-and-gold snappers* at Panama. (Author: LASZLO ILYES from Cleveland, Ohio, USA. Source: Snappers Galore. https://creativecommons.org/licenses/by/2.0/deed.en.)

FIGURE 10.2
School of big eye *scad* at Hawaii. (Author: Steve D. Source: Flickr, CC BY 2.0. https://creativecommons.org/li
censes/by/2.0/deed.en.)

FIGURE 10.3
Fish school. (Author: Matthew Hoelscher. Source: Flickr, CC BY-SA 2.0. https://creativecommons.org/licenses/
by-sa/2.0/deed.en.)

Basic swarm behavior is modeled on three simple rules:

- Move in the same direction as neighbors
- Avoid collisions with neighbors
- Remain in the vicinity (close to) of neighbors

Every entity of a swarm has multiple zones around itself. The sensory capabilities of the entity determine the zone size (neighborhood size) and shape of the zone. To avoid collision with neighbors, the entity maintains a minimum distance zone around itself. To align with the movement of its neighbors, it has a zone to sense the direction of movement and move along with them. Outside the periphery of these two zones the entity has a wider area to focus on other things based on its capabilities.

A shoal is a group of fish that normally belongs to the same species but could also contain a mixture of different species. A school describes a group of fish that often belong to the same species and swim together in a synchronized and polarized manner. This schooling behavior helps the fish in foraging for food and guarding themselves against predators and also in attracting mates. Fish prefer to form shoals with members that closely resemble each other in appearance since this increases their chance of protection against predators. Any fish that stands out among the group is a likely target for attacks. The members have an innate capacity to know in which direction to swim while foraging for food. Some of the species of fish migrate during certain periods of the year to different locations in huge numbers.

Fish select shoals that have a general appearance in size and shape similar to themselves. The larger the size of the shoal the better defense they have against predators, and the higher their chances of finding suitable mates and good-quality food. If any fish is out of place in the shoal, that is, it has an appearance or behavior that stands out from the rest of the shoal, it is more likely to be attacked by predators. Fish swim and forage independently, but they are aware of their neighbors and their place in the shoal, and they tend to keep in the vicinity of their neighbors while swimming and foraging. If the fish in the shoal all swim in the same direction and speed they turn into a school. When fishes are schooling they can even do complex maneuvers. They tend to be calm when they are within their shoals. They share information regarding the presence and location of food among members of the shoal. Ocean upswellings provide rich feeding grounds for fish. Smaller fish become feed for larger fish like sharks and whales. Fish schools are disciplined, and this schooling behavior is instinctively developed by the fish when they are young. Fish require good vision for exhibiting schooling behavior, and they also have other senses such as a lateral line running along their sides which helps in sensing neighbors.

The important parameters characterizing a shoal are size (number of fish), density (number of fish per volume), polarity (extent to which the fish are pointing in the same direction – average difference between the orientation of the group and the individual), nearest neighbor distance (distance between the centroid of one fish to another – related to density of the fish school), and nearest neighbor position (angle and distance between one fish to another). A school of fish with low polarity is shown in Figure 10.1. Schooling fish that are in high density are shown in Figures 10.2 and 10.3, whereas schooling fish that are in low density are shown in Figure 10.4. Fish prefer to keep together in shoals even though they might have to share the food.find and fall in line with the group discipline.

The complex behavioral patterns of fish within groups depend on the species, age, geographic location, environment, habitat, light levels, and other factors. The fish schools

FIGURE 10.4
Schooling *banner fish*. (Author: Jon Hanson from London, UK. Source: Schooling Bannerfish School. CC BY SA 2.0, https://creativecommons.org/licenses/by-sa/2.0/deed.en.)

are highly coordinated and tightly knit. Some of the schools extend several kilometers in length and several meters in width and depth. The shoal structure alters based on the current activity of the fish such as migration, foraging, and feeding. The size of the shoal also varies to a great extent, from hundreds of fish to millions of fish. They have an advantage of exploring more food patches compared to individual fish. Fish also migrate hundreds to thousands of kilometers, maintaining high speed. Isolated fish have higher stress leading to lower mobility, less adaptation to environment, and lesser exploratory capabilities. Living in groups curtails their movement and freedom to some extent and also food has to be shared, but the advantages of security against attacks and collective foraging outweigh the disadvantages. These characteristics and behavior of fish swarm have been incorporated into the fish school search optimization algorithm.

10.3 Fish School Search Optimization

Fish schools also follow the three simple rules of the swarm:

1. Move in the same direction as your neighbors.
2. Remain close to your neighbors.
3. Avoid collisions with neighbors.

The fish school search algorithm has been proposed for searching for optimum solutions in search spaces that have higher dimensions [2]. The FSS algorithm is mainly based on the three functions of fishes – *feeding*, *swimming*, and *breeding*. The three main operations of the algorithm are modeled on these three activities of the fish school. A detailed description of the three activities of fish on which this algorithm is based follows.

Feeding

Fish can eat food and grow strong or swim and lose weight. The search for food is equivalent to the search for candidate solutions in the search space. Better quality food is equivalent to better quality solutions (higher fitness values of the objective function) to the problem.

Swimming

It is an activity undertaken by all the fish in the swarm. It is a coordinated collective movement of the swarm of fish. When the fish are schooling they swim together in a coordinated manner oriented in the same direction, whereas if they are shoaling they swim together but with less coordination. Swimming is driven by feeding and guides the search for solutions.

Breeding

It is a part of the natural biological evolution process that takes place for producing the next-generation population. More fit individuals get selected for mating and breeding, and they produce offspring for the succeeding generations. The weaker individuals gradually get dropped from the population. These evolutionary mechanisms apply to the breeding process of fish. This is equivalent to producing solutions with higher fitness values in succeeding iterations.

In the FSS algorithm the agents that search for the solutions in the search space are the fish. FSS is suitable for unstructured search spaces with higher dimensions. Each fish could be a potential candidate solution for the problem to be solved. The weights of the fish represent their success in the search for optimal solutions; this is equivalent to memory in fish. Evolution is through a combination of operations such as swimming and breeding. The FSS algorithm is built upon the following principles:

- Computations are simple and within a small local area.
- Memorizing the result of past computations.
- Less communication between the fish.
- Distributed control (or less centralized control).
- Population diversity.

Incorporating these principles into the FSS algorithm simplifies the computations, thus reducing the computational complexity and also the time complexity, allows for information to be shared, and adapts learning into the search process, thus speeding up the convergence of the algorithm. In the FSS algorithm *aquarium* is the boundaries of the search space where the optimum solution could be found. The quantity of food is equivalent to the quality of solution at any point in the search space (for maximization problems). Food locations in the aquarium are promising locations where good candidate solutions are likely to be found. Swimming guides the search process of all the agents towards more promising regions of the search space within the aquarium. Breeding refines the search and switches between exploration and exploitation.

10.3.1 Algorithm

Let X_i^{iter} represent the position of the ith fish in iteration $iter$ (i = 1, 2, ..., N) where N is the population size of the swarm. Let $f(X_i) = f(x_{i1}, x_{i2}, ..., x_{id})$ represent the d-dimensional objective function evaluated at the position of the ith fish, and W_i^{iter} be the weight of the ith fish. The weight update of the fish in every iteration is given by Equation 10.1.

$$W_i^{iter+1} = W_i^{iter} + \frac{f(X_i^{iter+1}) - f(X_i^{iter})}{\max\{|f(X_i^{iter+1}) - f(X_i^{iter})|\}} \tag{10.1}$$

Fish swim in the water within the boundaries of the aquarium where food is available at various locations in differing quantities. Depending on the quantity of food at the current and next positions of the ith fish, the weights of the fish are updated. It is assumed that fish can vary in weight between a minimum of 1 and a maximum of W_{max} and on an average the initial weight of the fish is $W_{max}/2$.

Fish swim due to *individual, collective-instinctive*, and *collective-volitive* movements. Individually each fish moves (swims) in a particular direction if it assesses that the food in the new position (or direction) is better than at the current position. The fish moves in steps of s_i^{iter} which could be constant or could be adaptively changed with the increasing iterations. When exploration turns to exploitation, the step size could be decreased. To include some heuristics into this movement, the step size s_i^{iter} is multiplied by a random number r_i that takes on values in the interval [0, 1] with a uniform distribution. The fish moves in the chosen direction with the given step size provided that the new position is within the aquarium boundaries. This movement increases the exploitation ability of the algorithm.

When the fish of the school move individually, some are more successful than others in the quantity of food found in the new position. The average of the individual movement of all the fish in the school is computed. This gives the direction where the fish school has to orient itself for more successful foraging. Each fish takes up a new position based on this average direction in which the entire fish school has to move. The equation governing this *collective-instinctive* movement is given by Equation 10.2.

$$X_i^{iter+1} = X_i^{iter} + \frac{\sum_{i=1}^{N} \Delta X_i^{iter}\{f(X_i^{iter+1}) - f(X_i^{iter})\}}{\sum_{i=1}^{N}\{f(X_i^{iter+1}) - f(X_i^{iter})\}} \tag{10.2}$$

where ΔX_i^{iter} is the displacement of the ith fish in iteration $iter$, that is, the difference between the current and new position of the ith fish reached by taking a step size. The barycenter of the fish school is given by Equation 10.3.

$$B^{iter} = \frac{\sum_{i=1}^{N} X_i^{iter}.W_i^{iter}}{\sum_{i=1}^{N} X_i^{iter}} \tag{10.3}$$

The position of the fish is adjusted after *individual* and *collective-instinctive* movements according to the location of the barycenter of the fish school. Depending on whether the weight of the fish school has increased or decreased, the *collective-volitive* movement of the fish school will be either inward or outward respectively. This movement enhances the exploration capability of the FSS algorithm. For inward movement Equation 10.4 is applied whereas for outward movement Equation 10.5 is applied.

$$X_i^{iter+1} = X_i^{iter} - s_v.r_v.(X_i^{iter} - B_i^{iter}) \tag{10.4}$$

$$X_i^{iter+1} = X_i^{iter} + s_v r_v \left(X_i^{iter} - B_i^{iter} \right) \tag{10.5}$$

It is assumed that food is scattered all over the aquarium at different locations. The fish swim looking for food. They either grow in weight or diminish in weight depending on the food at the previous and current locations. The change in weight of the fish depends on the normalized difference in the food concentration at the previous and the current location. The food concentration represents the evaluation of fitness function at these locations.

Every fish has *individual* movement in each iteration. They swim in random directions looking for better sources of food within the boundaries of the aquarium. If the fish finds that the food lies outside the aquarium boundaries it does not exhibit movement. The weighted average of individual movements is computed. The fish that are successful have more influence on further movement. This computation of the weighted average determines the direction of movement in the next iteration. ΔX_i^{iter} is the displacement of fish i during the iteration *iter*. Each fish moves in a direction that is influenced by the weighted average computed. This is a collective evaluation of movement of all fish based on *individual* and *collective-instinctive* movements. If the fish is gaining weight that means it has found a better food source and the search region should contract; otherwise, the search region should dilate. With respect to the barycenter of the fish school, the fish take steps. The collective movement will be inwards or outwards depending on whether weight has been gained or has decreased.

Fish are selected for *breeding* based on a threshold. The fish that has maximum ratio of weight over distance with respect to the breeding fish is selected for the breeding operation. Let the parent fish be represented by X_p and X_q and the child fish by X_c. The equations governing the weight and position of the child fish are given by Equations 10.6 and 10.7.

$$W_c^{iter+1} = \frac{W_p^{iter} + W_q^{iter}}{2} \tag{10.6}$$

$$X_c^{iter+1} = \frac{X_p^{iter} + X_q^{iter}}{2} \tag{10.7}$$

When the new fish is created, the weakest (smallest fitness value) fish is removed from the population. The algorithm starts with an initial population of N fish with random positions (location) and weight in the search space (aquarium). The algorithm is iterative and converges when the global optimum is attained or a stopping criterion is met or a maximum number of iterations is reached. The operations defined above are repeated with every iteration until termination of the algorithm. The stopping criterion could be maximum time limit, maximum school weight, maximum school radius, maximum fish number, maximum breeding number, or any other defined criterion.

10.3.2 Pseudocode

Initialization

   Population size of fish swarm N

   Random position of the fish in the aquarium $X_i = [x_{i1} \ x_{i2} \ ... \ x_{id}], \quad i = 1, 2, ... N$

   Initial weights of fish $W_i \quad i = 1, 2, ... N$

   Objective function $f(X)$

   Define termination criteria, if any

   Maximum number of iterations *MaxIter*

   iter = 1

while (*iter* ≤ *MaxIter*) **do**

   for $i = 1$ *to* N **do**

      Evaluate fitness of fish

      Apply *individual movement* operator

      Apply *feeding* operator

      Evaluate fitness of fish

      Calculate weight of fish in new position

   end for

   for $j = 1$ *to* N **do**

      Apply *collective-instinctive movement* operator

   end for

   Compute barycenter of the fish school

   for $k = 1$ *to* N **do**

      Apply *collective-volitive movement* operator

   end for

   for $m = 1$ *to* N **do**

      Apply *breeding* operator

      Eliminate the weakest fish

   end for

end while

Fish with highest fitness value is the global optimum solution

Flowchart

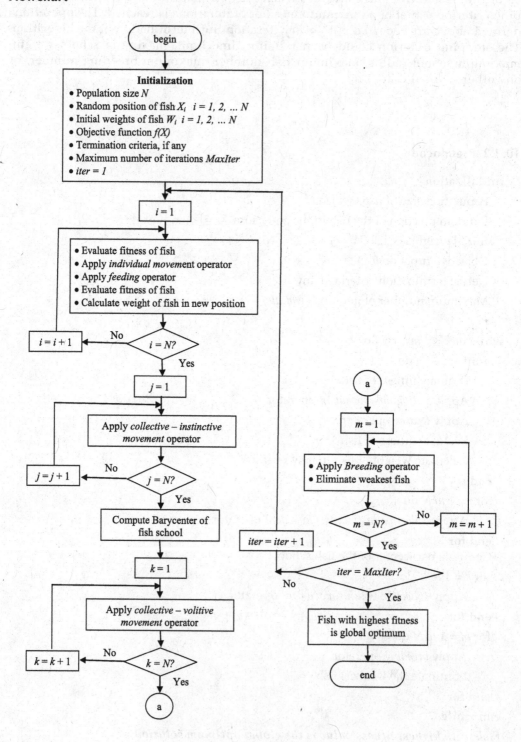

10.4 Variants and Applications

An improvement over the fish school search algorithm has been proposed by including particle swarm optimization and integrating communication within the algorithm [3]. FSA has higher global search capability and the fewest occurrences of getting trapped in local minima. The parameter setting complexity has been reduced, and its performance over evolutions has been studied. In [4] the artificial fish swarm algorithm has been applied to image registration and found to have good accuracy and speed, and also the convergence properties of the algorithm have been extensively studied. In engineering design, it is important to optimize the resources and produce the best output. Nature-inspired optimization algorithms play a central role in attaining the solutions to such complex problems. The FSS algorithm has been applied to standard engineering design problems and the results presented [5]. The weight-based FSS that was invented for multimodal problems has been modified as wrFSS and applied to solve multimodal constrained optimization problems [6]. Several sub-swarms have been employed to exploit the search space. Another improved version of the FSS namely fish school search – combined strategy (FSS–CS) has been proposed [7]. In this algorithm, feeding is enhanced, the exploration strategy is not greedy, and the steps in the movement operators have elliptic decay. The fitness values are used in the feeding operation rather than fitness variation as in the original algorithm.

In [8] the multi-objective FSS (MOFSS) has been proposed by modifying the FSS to solve problems with conflicting multiple objectives. The operators have been modified, and an external archive has been included to store non-dominated solutions. The MOFSS is meant to find the non-dominated solutions that lie on the Pareto Optimal Front. Several variants of the FSS have been proposed with enhancements that produce better results than the original FSS. One such version has been presented in [9] where there is one fitness evaluation for every fish in each iteration, and it has good exploitation properties and is simple to implement. In the enhanced algorithm, the operators are combined and information from earlier iterations is used in the current iteration. This enhanced FSS algorithm has been found to outperform the original FSS and PSO. There are also variants of the FSS for solving multi-objective and binary optimization problems. In addition the parameters of the FSS can be tuned and its impact on the algorithm performance can be studied. The algorithm can be extended to tackle problems with constraints and also modified for combinatorial optimization problems. Some of the constraints such as aquarium boundaries, maximum limits for weight of fish, fish school radius, and time complexity can be incorporated into the algorithm to refine the results.

10.5 Summary

A novel metaheuristic search algorithm based on the swarming behavior of fish has been discussed in this chapter. Fish schools are highly oriented for swimming, foraging, migration, and defense against predators. The complex interactions among the members within the fish swarm and the operations of feeding, swimming, and breeding have been inculcated in the development of the fish school search algorithm. FSS is a metaheuristic algorithm that is able to solve complex problems with simple mathematical models. The fish schools contain high volume and density in some species and could go up to thousands of fish. This makes them suitable for search in unstructured high-dimensional spaces for the optimum solution. The FSS algorithm also has its own specific characteristics and properties that make it suitable for high-dimensional, unstructured, multimodal search spaces.

FSS has a good balance between exploration and exploitation abilities. FSS produces excellent results for unimodal as well as multimodal NP-hard problems. On average, a population size of 20 to 50 with a maximum of 100 iterations is sufficient for solving most problems. Every nature-inspired algorithm has its own specialized features that are suited to a specific set of problems. FSS has been experimentally found to outperform PSO for standard benchmark data sets.

References

1. D. S. Pavlov, O. S. Kasumyan, Patterns and mechanisms of schooling behavior in fish: A review, *Journal of Ichthyology*, Vol. 40, Suppl. 2, pp. S163–S231, 2000.
2. Carmelo J. A. Bastos Filho, Fernando B. de Lima Neto, Anthony J. C. C. Lins, Antônio I. S. Nascimento, Marília P. Lima, A novel search algorithm based on fish school, behavior, *2008 IEEE International Conference on Systems, Man and Cybernetics (SMC 2008)*, pp. 2646–2651.
3. Hsing-Chih Tsai, Yong-Huang Lin, Modification of the fish swarm algorithm with particle swarm optimization formulation and communication behavior, *Applied Soft Computing*, Vol. 11, No. 8, pp. 5367–5374, 2011.
4. Yang Wang, Wei Zhang, Hongxing Li, Application of artificial fish swarm algorithm in image registration, *Computer Modeling and New Technologies*, Vol. 18, No. 12B, pp. 510–516, 2014.
5. Fran Sergio Lobato, Valder Steffen Jr., Fish swarm optimization algorithm applied to engineering system design, *Latin American Journal of Solids and Structures*, Vol. 11, No. 1, pp. 143–156, January 2014.
6. J. B. Monteiro-Filho, I. M. C. Albuquerque, F. B. Lima Neto, Fish school search algorithm for constrained optimization, *Neural and Evolutionary Computing*, pp. 1–12, November 2018.
7. Carmelo J. A. Bastos-Filho, Rodrigo P. Monteiro, Luiz F. V. Vercosa, Improving the performance of the fish school search algorithm, *International Journal of Swarm Intelligence Research*, Vol. 9, No. 4, pp. 21–46, October 2018.
8. Carmelo J. A. Bastos-Filho, Augusto C. S. Guimarães, Multiobjective fish school search, *International Journal of Swarm Intelligence Research*, Vol. 6, No. 1, pp. 23–40, 2015.
9. C. J. A. Bastos-Filho, D. O. Nascimento, An enhanced fish school search algorithm, *Proceedings of the 2013 BRICS Congress on Computational Intelligence and 11th Brazilian Congress on Computational Intelligence*, Washington, DC, United Sattes, pp. 152–157, 8–11 Septembr 2013.

11

Cuckoo Search Algorithm

11.1 Introduction

Metaheuristic algorithms are very powerful in solving complex real-life problems since they are inspired by natural phenomena. Biological organisms in nature have their own evolutionary processes and mechanisms of survival in hostile environments. The main characteristics of metaheuristic algorithms of intensification and diversification are responsible for the effective performance in solving complex engineering problems and NP-hard problems in computer science. Intensification and diversification lead to good exploitation and exploration properties of the algorithm which are necessary for intense local search and diversified global search. Inculcation of Levy flight behavior into the metaheuristic optimization algorithm improves the performance for exploitation as well as exploration. Some birds and insects, especially fruit flies, exhibit Levy flight behavior. Levy flight is movement of a bird or insect in straight-line paths punctuated by sharp 90° turns.

The cuckoo search (CS) optimization algorithm has been developed based on the brood parasitic behavior of cuckoo birds. It is a metaheuristic algorithm proposed by Xin-She Yang and Suash Deb in 2009 [1, 2]. Cuckoo birds are famous for the beautiful and musical sounds they produce which are pleasant to hear. Some species of cuckoo birds build their own nests and raise their young while many other species are brood parasites. The brood parasitism and Levy flight behavior are the two main traits of cuckoo birds that form the framework of the cuckoo search optimization algorithm. Every species in nature, including the cuckoo bird, instinctively follows Darwin's theory of *survival of the fittest*. They adapt themselves to the changing environmental conditions and exhibit swarm intelligence and properties as the manifestation of this adaptation. The cuckoo search algorithm has been built upon the swarm behavior and intelligence of cuckoo birds that have been described in the following sections.

11.2 Cuckoo Bird Behavior

The cuckoo belongs to the family of birds called *Cuculidae*. Cuckoo birds are of medium size, slender, and live in trees or on the ground. They have wide-ranging habitats. Some cuckoo species are found in tropical regions, especially in rain forests, whereas others are migratory. The migratory birds move during winter. Most of them live solitarily and feed on insects and fruits. Figure 11.1 shows a fan-tailed cuckoo found in Australia. Figure 11.2 shows a male Asian cuckoo called koel which is famous for its distinctive musical sounds, coo-coo.

FIGURE 11.1
Fan-tailed cuckoo. (By J.J. Harrison – own work, CC BY-SA 3.0. https://creativecommons.org/licenses/by-sa/3.0/deed.en.)

Cuckoos are brood parasites [3] and have soft feathers. *Cuculinae* is the subfamily of cuckoos that are brood parasitic. Cuckoos normally do not build their own nest, but some species do have their own nests and raise their young. The parasitic cuckoos are very poor parents. They use more than 100 species of other birds as their hosts. The parasitic cuckoos lay their eggs in the nest of other birds. The different species of cuckoos lay eggs that are white, grey, or colored such as green, red, and yellow, and they could be plain or spotted. These varieties of eggs can match the eggs of any host bird. Usually the cuckoo bird waits near the nest of a host bird watching for an opportunity to enter and lay its eggs. They try to mimic the eggs of the host bird so that their eggs will be preserved and hatched by the host bird. The cuckoos ensure that their eggs are hatched before those of the host bird. Cuckoo bird flies into the nest of a host bird, pushes out one of the eggs of the host bird and replaces it with its own egg. The female cuckoos visit different host nests and lay their eggs. The eggs almost match the host eggs in color and size, called egg mimicry. Figure 11.3 shows five eggs in a reed warbler nest including one cuckoo egg among the collection of four reed warbler (host) eggs, that matches in color and pattern but is slightly bigger.

The cuckoo chicks resemble those of the host bird so that their young ones get fed by the host. For example, in some species of cuckoos the host bird might be a crow and hence the cuckoo chick is black and resembles a young crow. Some of them inhabit reed beds where reed warblers have their nests and they become the host for the cuckoos. The cuckoo bird lays its eggs in the nest of the reed warbler, and once it is hatched, the cuckoo chick tries to push out the egg of the host bird. This is done to ensure that the cuckoo chick gets fed

FIGURE 11.2
Male Asian koel, Chalakudy, Kerala. (Author: Challiyan – own work, CC BY-SA 4.0/3.0/2.5/2.0/1.0. https://co
mmons.wikimedia.org/wiki/Commons:GNU_Free_Documentation_License,_version_1.2. https://creativ
ecommons.org/licenses/by-sa/4.0/.)

by the host reed warbler bird. A reed warbler feeding a common cuckoo chick in a nest is
shown in Figure 11.4.

Nests that are nearer to the vantage point of the cuckoos and that had more visibility
were more vulnerable to being hosts for the parasitic cuckoos. Cuckoo chicks persuade the
host birds to feed them by making begging calls. When the cuckoo chicks have outgrown
the nest, they fly out. Some species of cuckoos are obligate brood parasites, that is, they
reproduce only by using host nests. Some other species are non-obligate brood parasites,
that is, they lay their eggs in the nests of their own species of cuckoos and raise their young.

When the host bird discovers that the cuckoo bird has laid eggs in its nest, either the
eggs are evicted or the host abandons the nest and builds a new one. Some of the species
of cuckoos expertly time the laying of their eggs with the host bird and also mimic the
color and pattern (also size) of their eggs with the host, so that their own eggs will not be
discovered by the host. The cuckoo bird lays its own eggs in the host nest after eviction
of one egg of the host bird, and the cuckoo egg also hatches earlier. When it hatches, the
cuckoo chick tries to evict the eggs of the host bird so that it gets a good share of the feed
from the host bird. Different strategies are adopted by the cuckoos in carrying out brood
parasitism. As an example, the male cuckoo lures away the host bird from its nest so that
the female cuckoo can lays its egg in the host nest. Cuckoos are secretive birds, but they
have distinctive musical calls. Since cuckoos are mostly not raised by their own parents,
their behavior of evicting host eggs and calls are considered innate qualities.

FIGURE 11.3
Great reed warbler nest containing one cuckoo egg and four warbler eggs (Apaj, Hungary). (Author: Attila Marton 1 – own work, CC BY-SA 4.0. https://creativecommons.org/licenses/by-sa/4.0/deed.en.)

11.3 Levy Flights

Levy flights are named after the French mathematician Paul Levy. Some birds and insects, especially fruit flies, exhibit Levy flight behavior. They move in straight-line paths punctuated by sharp 90° turns. Stochastic processes are governed by random trajectories that are commonly known as random walk [4]. Levy flight is a particular class of random walk where the step length has a heavy-tailed probability distribution. Figure 11.5 shows an example of Levy flight movement with 1000 steps in a two-dimensional space.

Such Levy flight behavior of animals, birds, and insects has been found to be suitable for emulation in optimization algorithms. The probability density function of the Levy distribution is for continuous random variables that can take only positive values. It is given by Equation 11.1:

$$f(x,\mu,c) = \sqrt{\frac{c}{2\pi}} \frac{e^{\frac{-c}{2(x-\mu)}}}{(x-\mu)^{\frac{3}{2}}}$$

(11.1)

where μ is a location parameter that shifts the curve and c is a scale parameter. As $x \to \infty$ the *pdf* is approximately given by:

$$f(x,\mu,c) \approx \sqrt{\frac{c}{2\pi}} x^{-\frac{3}{2}}$$

(11.2)

The wing of the pdf is heavy and fat-tailed as given by Equation 11.2. Figure 11.6 shows the Levy distribution for $\mu = 0$ with different values of scale parameter. The Levy distribution has infinite mean and variance.

FIGURE 11.4
Eurasian reed warbler feeding a common cuckoo chick. (Author: Per Harald Olsen – own work, CC BY-SA 3.0. https://commons.wikimedia.org/wiki/Commons:GNU_Free_Documentation_License,_version_1.2, https://creativecommons.org/licenses/by-sa/3.0/deed.en.)

For large values of a random variable x (x could be the sum of several random variables) that follow Levy distribution, the probability density function (*pdf*) is approximately given by,

$$f(x) \approx x^{-(1+\mu)}, 0 < \mu < 2 \tag{11.3}$$

Random variables with infinite variance exhibit power law distribution with heavy and fat tails. Levy flights along with Brownian motion describe the pattern of movement of several animals and birds while foraging for food. Usually, animals and birds forage for food following a random walk. Since random walks can be modeled statistically, the location and transition probability determine the step size and direction of the movement. Levy flights have wide-ranging applications that include light behavior, animal foraging behavior, human walk, earthquake description, etc. It is shown that it is possible to develop optical materials that exhibit Levy flight behavior for light. This is a pioneering work that has made it possible to study Levy flight behavior under controlled conditions.

11.4 Cuckoo Search Optimization

Cuckoo search (CS) is a metaheuristic optimization algorithm based on the obligate parasitic breeding behavior of some species of cuckoo birds. It also inculcates the Levy flight

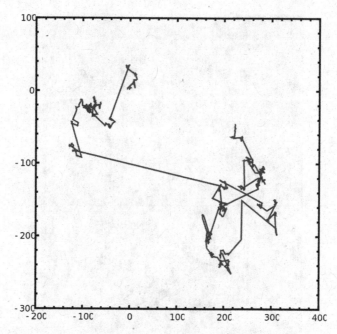

FIGURE 11.5
Two-dimensional Levy flight with 1000 steps. [By user: PAR – own work (Public Domain).]

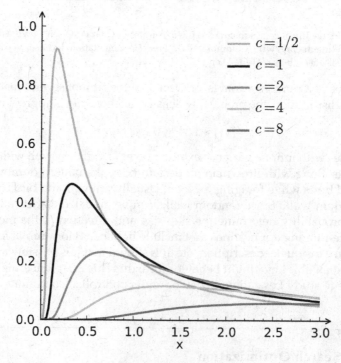

FIGURE 11.6
Levy distribution *pdf* for $\mu = 0$. (Author: Krishnavedala – own work, CC0 1.0. https://en.wikipedia.org/wiki/
en:Creative_Commons.)

movement exhibited by some animals, birds, and insects, including cuckoos. The cuckoo search optimization algorithm has been developed based on three assumptions:

- The number of host nests in which a cuckoo can lay its egg is fixed, represented by the population size N. Let $p_h \in [0, 1]$ be the probability that the host bird discovers the cuckoo egg laid in its nest. Once discovered, the host bird either evicts the egg or abandons the nest and builds a new one. Therefore, p_h is the fraction of the N nests that are either abandoned or replaced with new ones by the host bird. This characteristic is equivalent to abandoning solutions that have lesser fitness values and replacing them with new and better solutions. Each host nest represents one possible candidate solution to the problem.
- A cuckoo bird lays only one egg in a randomly chosen host nest at any instant of time.
- The nests with high-quality eggs have higher fitness values and hence survive to the next generation.

If the cuckoo egg is discovered by the host bird, the bird has two possible courses of action; either it can abandon the nest and build a new nest elsewhere, or it can throw out the egg of the cuckoo bird. Let p_h be the fraction of nests that are abandoned by the host, out of the total population of N nests and new nests are built in the next generation (replacing the abandoned ones). Each egg in a host nest (assuming only one egg is present) represents a solution to the problem, and the parasitic cuckoo egg represents a new solution which might replace the existing solution if it is found to be better.

An objective function is formulated as a mathematical expression based on the problem to be solved. The evaluation of the objective function with a set of input variables gives a possible solution to the problem. The algorithm randomly places the N host nests in the search space. For simplicity, it is assumed that there is only one host egg in the nest and the parasitic cuckoo lays one egg in the host nest. This is repeated for all the nests in the search space. The eggs represent possible solutions to the problem, and that means they are the evaluated fitness function values. If the fitness of the cuckoo egg is found to be better than the fitness of the host egg, the cuckoo egg replaces the host egg in the nest. Otherwise the cuckoo egg is thrown out of the nest. This is repeated for all the nests in the search space. In each iteration, the nests are updated based on their fitness values. At the end of the maximum number of iterations fixed for the algorithm, the nests with the worst fitness values are replaced with new ones.

11.4.1 Algorithm

Let X_k^{iter} represent the position of the *kth* nest of the host birds that are randomly located in the search space.

$$X_k^{iter} = [x_{k1}^{iter} \; x_{k2}^{iter} \; ... \; x_{kd}^{iter}] \quad k = 1, 2, ..., N \tag{11.4}$$

The search space is assumed to be d-dimensional, indexed by the variable j with $j = 1, 2, ..., d$. Therefore the objective or fitness function defined for the problem is d-dimensional given by

$$f(X) = f(x_1 \; x_2 \; ... \; x_d) \tag{11.5}$$

The iterations are indexed by the variable *iter*, with the maximum number of iterations being represented by *MaxIter*. If required, a stopping criterion could also be defined for the problem.

Let X_c^{iter} be the position of the *cth* cuckoo in iteration *iter* and X_c^{iter+1} be its new position in the next iteration. The position of the *cth* cuckoo in the succeeding iteration is governed by its current position and Levy flight movement as given below:

$$X_c^{iter+1} = X_c^{iter} + s \oplus Levy(u) \tag{11.6}$$

where s is the step size that is always positive and the value depends on the problem; usually it can be chosen as 1. *Levy(u)* is the Levy distribution that models the transition probability, thus making the next position depend on the present position and the transition probability.

$$Levy(u) \approx t^{-u}, \; 1 \le u \le 3 \tag{11.7}$$

This random walk using Levy flight is efficient in searching for the optimum since the step size can be made longer or shorter. The new solutions are generated by random walk with a large step size for better exploration of the search space, and short step size for better exploitation of local regions in the search space. When the step size is large the algorithm can jump out of local optimum and reduces the possibility of being trapped in locally optimum solutions. A sequence of such random walks modeled using Levy flight behavior becomes a Markov chain. Levy flights are for global search, maintaining the diversity of the population, and for random walks to intensify the search. Levy flights have long trajectories for global search interspersed with short Brownian motion for intense local search. The probability parameter p_h balances the search between local and global. When $p_h = 0.25$, the search is 75% global and 25% local. The dynamic step sizes possible in Levy flights make the search efficient since it can adapt between diversification and intensification.

The *cth* cuckoo and the *kth* host nest are chosen randomly and their fitness evaluated. On comparing their fitness values, if the fitness of the cuckoo is greater than that of the host, the cuckoo egg replaces the egg of the host bird in the chosen nest. Otherwise, the host egg is not disturbed. The fraction p_h of the nests with the worst fitness values is abandoned. The nests are ranked according to their fitness values, and at the end of the maximum number of iterations, the nest with the highest fitness is the global optimum solution.

11.4.2 Pseudocode

Initialization

    Create host nests in the search space randomly with population size N

    Define d-dimensional objective function $f(X)$, $X = \{x_1, x_2, ..., x_d\}$

    Define stopping criteria, if any

    Maximum number of iterations *MaxIter*

    iter = 1

while (*iter* $\le$ *MaxIter*) **do**

    Randomly choose a cuckoo X_c and evaluate its fitness value F_c

    Randomly choose a host nest k and evaluate its fitness F_k

    if ($F_c > F_k$)

      replace the host egg with the cuckoo egg

    end if

    Fraction p_h of host nests with least fitness values are discarded

    Nests with higher fitness are retained and new ones built to replace the discarded ones

    All the nests (solutions) are ranked according to their fitness values

Nest with highest fitness chosen as the current best optimal solution to the problem

if *stopping criteria met* **exit**

else *continue*

iter = iter + 1

end while

Nest with best fitness value is the global optimum solution

Flowchart

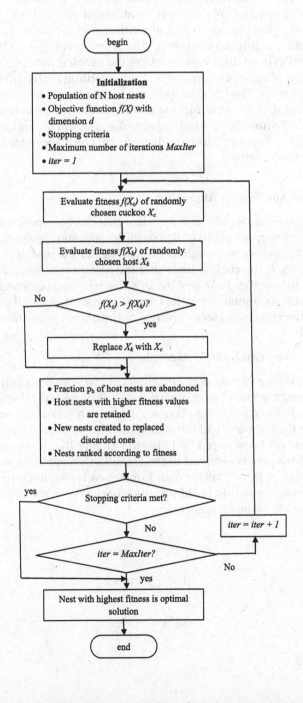

begin

Initialization
- Population of N host nests
- Objective function $f(X)$ with dimension d
- Stopping criteria
- Maximum number of iterations *MaxIter*
- *iter = 1*

Evaluate fitness $f(X_c)$ of randomly chosen cuckoo X_c

Evaluate fitness $f(X_k)$ of randomly chosen host X_k

$f(X_c) > f(X_k)$? — No

yes

Replace X_k with X_c

- Fraction p_h of host nests are abandoned
- Host nests with higher fitness values are retained
- New nests created to replaced discarded ones
- Nests ranked according to fitness

Stopping criteria met? — yes

No

iter = MaxIter? — No

iter = iter + 1

yes

Nest with highest fitness is optimal solution

end

11.5 Variants of the Algorithm

Since the original cuckoo search algorithm was proposed in 2009, several variants of the algorithm have been developed by various researchers working in the area of swarm intelligence [5]. These variants either modify the algorithm to be more efficient or make it adaptable for diverse applications. Some of the notable variants of the algorithm are modified CS, parallelized CS, binary CS, discrete CS, neural-based CS, and multi-objective CS. For some problems which are difficult to solve, especially continuous optimization problems, the optimal solution can be obtained efficiently by combining the CS algorithm with some other swarm-based algorithm such as GA or bat algorithm so that a hybrid optimization algorithm evolves to find a better solution to the problem. Multi-objective optimization is another variant where the problem consists of conflicting, multiple objectives. A set of solutions forming a Pareto Optimal Front will be suitable for such multi-objective problems. The CS algorithm has been adapted to solve specific engineering design problems such as wind turbine blades, steel truss structures, antenna arrays, optimal sequence attainment, optimal capacitance placement, and also NP-hard problems such as TSP, job scheduling, and graph coloring. The three notable variants of the CS algorithm are discussed below.

11.5.1 Discrete Cuckoo Search Algorithm

The discrete cuckoo search algorithm is a variant that has been proposed [6] to solve combinatorial optimization problems like the traveling salesman problem. It utilizes the fact that the cuckoos are intelligent enough to assess whether the host nest is suitable for their eggs and their chicks. If the cuckoo finds that the host nest is unsuitable for its parasitic breeding, it looks for another host nest. In this discrete cuckoo search algorithm a new class of cuckoos with such intelligence has been incorporated into the original algorithm to improve the performance and solve discrete optimization problems.

11.5.2 Binary Cuckoo Search (BCS) Algorithm

The BCS algorithm [7] is a variant of the original cuckoo search algorithm with the values assumed by the design vectors being binary, 0 or 1. It has a typical application to feature selection problem. Feature selection plays a vital role in classifiers since the number of features influences the complexity of the algorithm. The higher the number of features, the more computations will be required and also the greater the time complexity. The selection of appropriate features is essential in the classification accuracy, and this makes it an optimization problem. A binary vector with 1 or 0 in each position represents whether the feature will be selected or not. To build up this binary vector, the following modification in the equation is made in the CS algorithm:

$$X_k^{iter+1} = \begin{cases} 1 & if \quad P(X_k^{iter}) > \sigma \\ 0 & otherwise \end{cases} \tag{11.8}$$

$$P(X_k^{iter}) = \frac{1}{1+e^{-X_k^{iter}}} \tag{11.9}$$

11.5.3 Multi-Objective Cuckoo Search Algorithm (MOCS)

Most of the complex engineering design problems involve optimization of more than one conflicting objective with or without design constraints. When there are no constraints it is an unconstrained optimization problem. When there are one or more constraints on the design or solution, it is a constrained optimization problem. Moreover, the constraints can be linear or non-linear. For single-objective problems there is only one global optimum although there may be multiple local optima. For multi-objective problems, there are multiple solutions that involve a tradeoff between conflicting design objectives and constraints. The various possible optimal solutions form a Pareto Optimal Front, and any point on the front is a good enough solution to the problem. When one criterion is satisfied, the other ones might not be satisfied as much. There is no single solution that could become the global optimum for the entire set of objectives and constraints. It is effectively a tradeoff between the different objectives and constraints. One possible way of making the existing single objective optimization algorithms work for multiple objectives is to combine the multiple objectives as a weighted sum into a single objective. The search space should be sampled or searched such that there is enough diversity in the solutions obtained.

The CS algorithm is modified to extend it for optimization of multiple objectives in the problem [8, 9]. This becomes a multi-objective cuckoo search algorithm (MOCS). In the assumptions made in the original single-objective CS algorithm, each cuckoo was assumed to lay one egg in the host nest. In MOCS, each cuckoo lays K number of eggs in the host nest where each egg is a solution to one of the multiple objectives. Another assumption was the probability of discovery of a cuckoo egg in a nest by a host bird is p_h and the host either throws out the cuckoo egg or abandons its old nest and builds a new nest. In MOCS, the same assumption is made except that the host builds a new nest with K eggs. The solutions obtained for the multi-objective optimization problem lie on the Pareto Optimal Front and are non-dominated optimal solutions.

The pseudocode of the CS algorithm can be modified for MOCS as follows:

Initialization

    Create population of N host nests (each with K eggs)

    Define multiple objective functions $\{f(X_k)\}$, $k = 1, 2, \ldots, K$, $X = \{x_1, x_2, \ldots, x_d\}$

    Define stopping criteria

    Maximum number of iterations *MaxIter*

    iter = 1

while (*iter* ≤ *MaxIter*) **do**

    Randomly choose a cuckoo X_c and evaluate its fitness value F_c

    Randomly choose a host nest k and evaluate its fitness F_k for $k = 1, 2, \ldots, K$

    if ($F_c > F_k$)

      replace the host egg with the cuckoo egg

      <this step is done for all the eggs in the nest>

    end if

    Fraction p_h of host nests are abandoned based on their fitness values (those with least fitness are discarded)

    Nests with higher fitness values are retained and new ones are built to replace the discarded host nests

All the solutions (nests) are ranked according to their fitness values and the nests with fitness values that lie on the Pareto Optimal Front are chosen

if *stopping criteria met* **exit**

else *continue*

iter = iter + 1

end while

Nests with fitness values on the Pareto Optimal Front form the global optimum solutions

The solutions are Pareto Optimal or non-dominated if no other solution can be found that is better than the current solution. The set of all non-dominated solutions forms the Pareto Optimal Front. MOCS is efficient for multi-objective optimization problems. The algorithm has been validated for multi-objective test functions and engineering design problems with multiple, conflicting objectives and constraints.

11.6 Summary

The cuckoo search algorithm is based on the obligate parasitic breeding behavior of cuckoo birds combined with Levy flights. The Levy flights model the random walk that follows a Levy distribution with a heavy tail. The stochasticity in the algorithm is created by the Levy flight movement of cuckoo birds. The population size N and the probability p_h are two parameters of the cuckoo search algorithm that need to be chosen for the problem, and it has been found that the performance of the algorithm can be tuned by the appropriate choice of these two parameters. Hence $N = 15$ and $p_h = 0.25$ is sufficient for most of the problems. The cuckoo search algorithm has local as well as global search capabilities and can home in on the global optimum for unimodal as well as multimodal functions. As the algorithm runs, the nests aggregate at the global optimum for unimodal functions, and when the function is multimodal, the nests distribute themselves at positions of the local optima as well as global optima. Thus, cuckoo search finds all the local optima simultaneously provided the number of nests is much more than the number of local optima. The convergence rate of the algorithm does not depend on the above parameters. This makes it suitable for a wide variety of problems including NP-hard, single- and multi-objective problems.

This algorithm can be applied as it is or it can be hybridized with other metaheuristic algorithms such as GA and PSO. Compared to PSO and GA, the cuckoo search algorithm outperforms them for all benchmark unimodal and multimodal test functions. It is superior because it is able to have diversity of solutions and is also able to intensify the search in local regions by means of Levy flights. It also discards the solutions that are not good in every iteration and replaces them with better solutions. Cuckoo search is robust and applicable across a wide spectrum of problems. The CS algorithm has wide-ranging applications from feature selection, face recognition, thresholding, forecasting, and neural networks to the intractable problems in the computer science domain. The CS algorithm hybridized with other nature-inspired algorithms and its variants has been found to be more powerful in solving tough optimization problems.

References

1. X.-S. Yang, S. Deb, Engineering optimisation by cuckoo search, *International Journal of Mathematical Modelling and Numerical Optimisation*, Vol. 1, No. 4, pp. 330–343, 2010.
2. X.-S. Yang, S. Deb, Cuckoo search via Lévy flights, In: *Proceedings of World Congress on Nature and Biologically Inspired Computing (NaBIC 2009)Figu*, Coimbatore, India, pp. 210–214, December 2009, published by IEEE, USA, ISBN: 978-1-4244-5053-4.
3. Iztok Fister Jr., Dusan Fister, Iztok Fister, A comprehensive review of cuckoo search: Variants and hybrids, *International Journal of Mathematical Modelling and Numerical Optimisation*, Vol. 4, No. 4, pp. 387–409, 2013.
4. P. Barthelemy, J. Bertolotti, D. S. Wiersma, A Lévy flight for light, *Nature*, Vol. 453, pp. 495–498, 2008.
5. I. Fister Jr., X. S. Yang, D. Fister, I. Fister, Cuckoo search: A brief literature review, In: *Cuckoo Search and Firefly Algorithm: Theory and Applications, Studies in Computational Intelligence*, Vol. 516, pp. 49–62, 2014.
6. Aziz Quaarab, Belaid Ahiod, Xin-She Yang, Discrete cuckoo search algorithm for the traveling salesman problem, *Neural Computing and Applications*, Vol. 24, No. 7–8, pp. 1659–1669, June 2014.
7. L. A. M. Pereira, D. Rodrigues, T. N. S. Almeida, C. C. O. Ramos, A. N. Souza, X.-S. Yang, J. P. Papa, A binary cuckoo search and its application for feature selection, *2013 IEEE International Symposium on Circuits and Systems (ISCAS2013)*.
8. Waleed Yamany, Nashwa El-Bendary, Aboul Ella Hassanien, Eid Emary, Multi-objective cuckoo search optimization for dimensionality reduction, *20th International Conference on Knowledge Based and Intelligent Information and Engineering Systems, Procedia* Computer Science (Elsevier), Vol. 96, pp. 207–215, 2016.
9. X.-S. Yang, S. Deb, Multiobjective cuckoo search for design optimization, *Computers and Operations Research*, Vol. 40, pp. 1616–1624, 2013.

12

Firefly Algorithm

12.1 Introduction

Nature-inspired metaheuristic algorithms are mainly developed from the study of animals, birds, and insects, while some of them are derived from evolutionary strategies as well as physical and chemical processes. They are powerful as well as simple enough to provide solutions to complex problems effectively, with reduced time complexity. These algorithms are modeled on the characteristics of animals, birds, and insects and mimic their behavior. The mathematical models for these nature-inspired algorithms are built, and solving the related equations using appropriate techniques produces optimum solutions for the problems. The bioinspired algorithms incorporate metaheuristics and are simple to implement. Many of them are suitable for continuous and discrete variables, unimodal as well as multimodal problems. The objective function is defined based on the problem and may or may not depend on the landscape of the search space.

One such nature-inspired algorithm is the firefly algorithm (FA) that is modeled on the characteristics and behavior of fireflies in nature. Fireflies produce flashing light and are quite an awesome sight in the sky during the night. They normally fly around in groups and are a sight to behold with their brilliant flashing colors. They attract insects and other fireflies by their flashing lights for trapping prey as well as for mating. The flashing light of fireflies is rhythmic with a particular rate and pattern that is used in communicating with other members of the species. The fireflies vary in size, shape, color and rate of flashing, and the wavelength of light emitted. They are very sensitive to surrounding lighting conditions and to changes in the environment. They also exhibit Levy flight behavior which is flying in straight-line paths punctuated by sharp 90° turns. This intermittent scale-free search movement is exhibited by several species of birds and insects and is one of the main factors responsible for global search. The FA is population-based, and multiple fireflies (agents) search in the space in parallel. This inherent parallelism is one of the important factors in the efficiency of the algorithm in attaining the global optimum solution.

12.2 Firefly Behavior and Characteristics

Fireflies are insects that belong to the family Lampyridae, are soft-bodied, and are in the beetle order *Coleoptera*. They are found in tropical regions, mainly in South-East Asia, and there are more than 2000 species of fireflies, commonly called *lightning bugs*. Fireflies exhibit flashing behavior during twilight and during nighttime. They rhythmically flash light called bioluminescence to attract mates and prey, but the light does not have any ultraviolet or infrared radiation. Bioluminescence is a chemical reaction that takes place in the lower abdominal portion of the body of the firefly. The lower portion of the body of the fireflies radiates green, yellow, or a light shade of red, sometimes even blue light, mostly in the wavelength range of 510 to 670 nm. They are soft-bodied beetles with leathery wings and are approximately one inch in length. Figure 12.1 shows an adult beetle that is commonly known as a *firefly* or *lightning bug*. Figure 12.2 shows a glowing firefly that is emitting light from the lower portion of the body.

The flashing behavior, especially the color of the flashing lights, and the group size of the fireflies vary from species to species. The fireflies also exhibit variation in features such as size, shape, and color among the members of their family. The male fireflies exhibit synchronized flashing in large groups in order to attract the females. In some species the fireflies eat others who are attracted by the flashing lights. The fireflies themselves can be quite poisonous to other insects or vertebrates. They are also known as lightning bugs since they flash light to attract mates as well as prey. Figure 12.3 shows a female firefly in the grass that is emitting light.

The flashing rate and the duration vary from species to species and attract suitable mates. The rhythmic flashing not only attracts sexual partners but also unsuspecting prey. In some species, the males flash lights rhythmically to attract the females, whereas in some other species, the females flash light rhythmically (synchronized flashes), attract males, and eat them. These fireflies (fireflies are sometimes called glowworms; glowworms do not fly, but fireflies do fly) produce light from their bodies due to a chemical reaction (chemicals secreted from the abdomen of fireflies) which could be from parts of the body other than the abdomen such as the tail. The chemical energy is converted to light energy very efficiently (nearly 100%). Fireflies are distasteful to their predators, but they eat worms and bugs that live upon or under the ground, and when they grow up and learn to fly, they feed on other insects, nectar, and pollen in flowers.

Fireflies have organs under the abdomen that produce a chemical called *luciferin* that reacts with oxygen in the presence of the enzyme *luciferase* to produce light, a phenomenon

FIGURE 12.1
Adult beetle in the family Lampyridae, commonly called firefly. (Author: Bruce Marlin – own work, CC BY-SA 2.5. https://creativecommons.org/licenses/by-sa/2.5/deed.en.)

FIGURE 12.2
Glowing Eastern USA firefly (*Photinus pyralis*). (Source: Art farmer from Evansville, Indian, USA, CC BY-SA 2.0. https://creativecommons.org/licenses/by-sa/2.0/deed.en.)

called *bioluminescence*. Almost the entire energy of the chemical reaction is converted to light with no energy wastage as heat or in any other form. The fireflies are in larvae form when they are born and live under the ground till they become adult and grow wings to fly. The larvae also exhibit bioluminescence, but the flying adults are the ones that perform the dance with flashing lights. Fireflies are very sensitive to the light in their surroundings. The population of fireflies has been dwindling in recent years due to the use of

FIGURE 12.3
Common glowworm (*Lampyris noctiluca*), Aston, UK. (Author: Timo Newton-Syms – CC BY SA 2.0, https://creativecommons.org/licenses/by-sa/2.0/deed.en.)

FIGURE 12.4
Flashing fireflies in the forest near Nuremberg, Germany. (Author: Quit007, CC BY-SA 3.0. https://commons
.wikimedia.org/wiki/Commons:GNU_Free_Documentation_License,_version_1.2; https://creativecommons.o
rg/licenses/by-sa/3.0/deed.en.)

pesticides and changes in climate, environment, and other detrimental factors. Figure 12.4
shows flashing fireflies in the woods near Nuremberg, Germany.

Based on the characteristics and flashing behavior of fireflies, the firefly optimization
algorithm was developed by Xin-She Yang in 2009. The firefly optimization algorithm, its
variants, and applications have been described in the following sections.

12.3 Firefly-Inspired Optimization

The firefly optimization algorithm is a metaheuristic algorithm that is inspired by the
flashing behavior and Levy flight movement of fireflies [1]. The firefly optimization algorithm has been developed with the following three simplifying assumptions:

- The fireflies are unisexual so that any firefly will be attracted to any other firefly.
- The attractiveness of the fireflies is directly proportional to the brightness of light
  emitted by them. Because of this attractiveness, one firefly will move towards
  another one with a higher brightness. If there is no other firefly with higher brightness than its own, then it will move randomly.
- If the objective function is defined for maximization, then its value is proportional
  to the brightness of a firefly.

Let the population of N fireflies be distributed randomly in the d-dimensional search space. Let X_i represent the position of the *ith* firefly in the d-dimensional search space with $X_i = [x_{i1}\ x_{i2}\quad x_{id}]\ i = 1, 2, \ldots, N$. The objective function is defined as $f(X)$ where the fitness or objective function value is proportional to the light intensity of the firefly, that is, $I_L(X_i) \propto f(X_i)$. Light intensity decreases in inverse proportion to the square of the distance from the light source given by Equation 12.1.

$$I_L(m) \propto \frac{1}{m^2} \tag{12.1}$$

where $I_L(m)$ is the received light intensity at distance m from the light source. The diminishing light intensity with distance makes the flashing light of fireflies visible for a few hundred meters at night. This is sufficient for other fireflies to be attracted by the dancing rhythmic lights. The attractiveness is directly proportional to the brightness of light emitted and inversely proportional to the distance from the light source, as given in Equation 12.2 (inverse square law).

$$I_L(m) = \frac{I_o}{m^2} \tag{12.2}$$

where I_o is the original light intensity at source and m is the distance separating the source and point of reception (or point of measurement of light intensity). This decrease in intensity not only depends on distance but also on the absorption of the medium (typically air) in between. If the medium in which light travels has a fixed light absorption coefficient λ, then the light intensity variation with distance can be modeled as

$$I_L(m) = I_o e^{-\lambda m} \tag{12.3}$$

In Equation 12.2 the intensity becomes undefined at $m = 0$ (point of singularity). Hence the light intensity is modeled by including the inverse square law and light absorption; thus combining Equations 12.2 and 12.3 leads to Equation 12.4.

$$I_L(m) = I_o e^{-\lambda m^2} \tag{12.4}$$

If the rate of decay of light intensity is to be decreased monotonically, the following expression can be used:

$$I_L(m) = \frac{I_o}{1 + \lambda m^2} \tag{12.5}$$

Using Taylor's series expansion for exponential functions:

$$e^{-\lambda m^2} \approx 1 - \lambda m^2 + \frac{1}{2!}\lambda^2 m^4 - \ldots \tag{12.6}$$

$$\frac{1}{1 + \lambda m^2} \approx 1 - \lambda m^2 + \lambda^2 m^4 - \ldots \tag{12.7}$$

If the series expansion in Equations 12.6 and 12.7 is truncated to the first two terms, they can be equated as:

$$e^{-\lambda m^2} \approx \frac{1}{1+\lambda m^2} \approx 1-\lambda m^2 \qquad (12.8)$$

Since calculating $\dfrac{1}{1+\lambda m^2}$ is faster than computing $e^{-\lambda m^2}$ the attractiveness of the light intensity can be made proportional to $\dfrac{1}{1+\lambda m^2}$ as given in Equation 12.5. Let the attractiveness of the firefly be proportional to the light intensity and be represented by A_i for the *ith* firefly. Then the variation of attractiveness with distance is given by

$$A(m) = \frac{A_o}{1+\lambda m^2} \qquad (12.9)$$

In the above equation $e^{-\lambda m^2}$ becomes e^{-1} when $m = \dfrac{1}{\sqrt{\lambda}}$. Therefore the distance $m = \dfrac{1}{\sqrt{\lambda}}$ is called the characteristic distance m_C. Over the characteristic distance the attractiveness changes from A_o to $A_o e^{-1}$. In general, the attractiveness can be a monotonically decreasing function with distance such as $A(m) = A_o e^{-\lambda m^r}$ where $r \geq 1$. If the absorption coefficient λ is fixed, the characteristic length or distance is $m_C = \dfrac{1}{\lambda^{\frac{1}{r}}}$. As $r \to \infty$, $m_C \to 1$. If m_C is fixed, the parameter $\lambda = \dfrac{1}{m_C^r}$.

12.3.1 Algorithm

The Cartesian distance between any two fireflies X_j and X_k is $D_{j,k} = ||X_j - X_k||$. This distance is defined as $D_{j,k} = \sqrt{\sum_{n=1}^{d}(x_{jn} - x_{kn})^2}$ where $X_j = [x_{j1}\ x_{j2}\ ...\ x_{jd}]$ and $X_k = [x_{k1}\ x_{k2}\ ...\ x_{kd}]$.

The movement of the *jth* firefly towards the *kth* firefly that has higher brightness is given by

$$X_j^{new} = X_j + A_o e^{-\lambda m_{jk}^2}(X_k - X_j) + c.(rand - (1/2)) \qquad (12.10)$$

In the Levy-flight firefly algorithm (LFA) the movement of the firefly is modeled [2] using Equation 12.11.

$$X_j^{new} = X_j + A_o e^{-\lambda m_{jk}^2}(X_k - X_j) + c.sign(rand - (1/2)) \oplus Levy(u) \qquad (12.11)$$

The second term is due to the attraction between the *jth* and the *kth* fireflies, and the third term is for introducing randomness into the movement of the fireflies. c is the parameter introducing stochasticity into the movement of the fireflies, *rand* is a random number that assumes values in the interval [0, 1] with a uniform distribution, and *Levy(u)* is the Levy flight movement parameter. For most of the applications A_o can take the value of 1, and c can take any value in the range [0, 1]. *Levy(u)* is the Levy distribution that models the transition probability, thus making the next position depend on the present position and the transition probability.

$$Levy(u) \approx t^{-u}, \quad 1 \leq u \leq 3 \tag{12.12}$$

Levy distribution has infinite mean and infinite variance with a heavy tail.

When $\lambda \rightarrow 0$, the attractiveness $A = A_o$ is constant, meaning the flashing light of the firefly could be seen everywhere with the same intensity. This is a special case of PSO. When $\lambda \rightarrow \infty$, $m_C \rightarrow 0$ and $A(m) \rightarrow \delta(m)$ where $\delta(m)$ is the *Dirac delta* function. In other words, the attractiveness is almost zero for other fireflies, and they roam or move around randomly. This makes it a random search. The firefly algorithm lies between these two extremes, and the efficiency depends on the chosen parameters. In the different dimensions if the scales vary widely, for example, 10^5 in one dimension and 0.001 in another dimension, c can be scaled by the multiplication factor *scale(n)*, thus making it *c.scale(n)*, where $n = 1, 2, ..., d$.

12.3.2 Pseudocode

Initialization

    Population size of fireflies N

    Randomly locate the fireflies in the d-dimensional search space X_i, $i = 1, 2, ..., N$
        where $X_i = [x_{i1} \ x_{i2} \ ... \ x_{id}]$

    Define the objective function $f(X_i) = f(x_{i1}, x_{i2}, ... x_{id})$

    Light intensity of all the fireflies is determined by evaluating $f(X_i)$, $i = 1, 2, ..., N$

    Light absorption coefficient λ

    Stopping criteria, if any

    Maximum number of iterations *MaxIter*

    iter = 1

while (*iter* $\leq$ *MaxIter*) **do**

    for $j = 1$ to N

      for $k = 1$ to N

        if $(I_j > I_k)$

          Move firefly k towards firefly j

        end if

          Update light intensity of the fireflies at the new positions

      end for

    end for

    Rank the fireflies according to fitness and find the *current best*

    if stopping criteria met **exit, else** continue

    iter = *iter* + 1

end while

Firefly with the highest fitness value is the global optimum solution

Flowchart

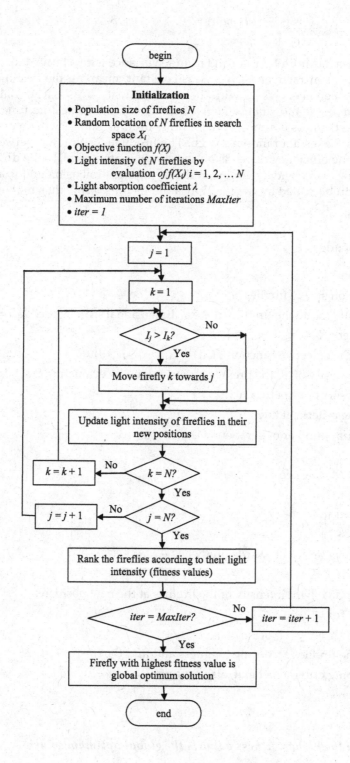

12.4 Variants and Applications

The firefly algorithm is suitable for non-linear, unimodal, and multimodal optimization problems and is found to be more efficient than GA and PSO. FA is also effective in solving multi-objective optimization problems. Discrete versions of the firefly algorithm have been proposed and are available in the literature with proven good performance. Hybrid optimization algorithms where the FA has been applied in combination with other nature-inspired algorithms have also been proposed. In [3] the firefly algorithm has been applied for the optimization of queueing systems that are used for the analysis and solution of complex problems related to the field of computer science as well as in industries. Vector quantization is a popular technique for image compression. The Linde–Buzo–Gray (LGB) algorithm is normally used to construct the codebook for VQ. In [4] the firefly algorithm has been applied along with the LGB algorithm to construct the optimal codebook. The optimal codebook design is one which maximizes the fitness function for all input vectors. The firefly-LBG algorithm has been compared with other state-of-the-art algorithms, and it has been found to outperform the other optimization algorithms. FA has been found to be suitable for clustering, image classification, feature selection, and for other computer science applications such as graph coloring, network routing, and the famous traveling salesman problem.

The fireflies have a natural habit of dividing themselves into groups, and, relating this behavior with our algorithm, it leads to grouping of fireflies around local optima. As the iterations progress, the groups cluster more and more around the optimum regions in the landscape and among them the global optimum can be identified. The parameter tuning gives flexibility in altering the performance of the algorithm as suited for the applications. The change in performance of the algorithm by varying the parameters has been explored [5]. The intermittent search strategy is one of the key components applied here where there is exploration of the landscape using Levy flights and intense exploitation of the search around regions of optimality [6]. Local search concentrates around regions where the global optimum is likely to be found. More exploration requires more iterations and hence convergence occurs later, whereas more exploitation requires fewer iterations which may lead to the global optimum or it could lead to premature convergence at a local optimum point. The idealized rules of the firefly algorithm have been combined with Levy flights to form the Levy-flight firefly algorithm (LFA) in [2]. There is a vast literature available on the firefly algorithm, its variants, and applications with hundreds of papers being published. In this chapter, a few variants and applications of FA have been discussed.

12.5 Summary

The firefly algorithm has been modeled on the flashing behavior of fireflies which is unique to their species. The fireflies are insects that exhibit Levy flight behavior. The Levy flights are used in inculcating stochasticity and diversity in the search for the global optimum solution. The flashing lights create attractiveness amongst the fireflies, and they group together. This grouping behavior is responsible for intense local search around local as well as global optima. The flashing light of the fireflies is associated with the objective function to be optimized. The brightness of a firefly at any location in the search space determines the objective function value at that location. Therefore the quality of the solution attained is directly proportional to the brightness of the light emitted by the firefly at the global optimum point.

Initially, the population of fireflies is distributed uniformly over the entire search space, and, normally within 50 to 100 iterations, convergence takes place. The typical population size suitable for most of the applications is 10 to 50. The greater the number of fireflies, the faster will be the convergence. The fireflies are almost independent within the swarm, thus their activities take place in parallel, increasing their efficiency. For most of the problems, experimentally it has been found that the parameters of the algorithm can be chosen as $c \in [0, 1]$, $A_o = 1$, $\lambda = 1$, $u = 1.5$. The change in attractiveness of the firefly with distance is characterized by the parameter λ, and it determines the speed of convergence of the algorithm. The attractiveness parameter λ typically varies from 0.01 to 100.

The Levy flight behavior of fireflies makes it possible to diversify the search on a global scale. The firefly algorithm with Levy flights has been found to outperform GA and PSO [7]. Like PSO, FA is also based on swarm intelligence, and PSO is a special case of the FA for certain settings of the parameters. FA is effective in solving intractable problems that have been found to be NP-hard for conventional algorithms. It has reduced time and computational complexity with few parameters to be tuned. The firefly algorithm has become very popular because of the promising results it produces for complex engineering problems. It has demonstrated superiority over other algorithms because of its balance between intensification and diversification. The Levy flight movement enhances the stochastic global search. The grouping of fireflies due to their attractiveness with each other leads to intense local search. The random factor in the algorithm follows uniform or Gaussian distribution, as the case may be. FA is a swarm intelligence optimization algorithm since it is based on the swarm behavior of fireflies and hence exhibits the characteristics and properties of other swarm-based algorithms in nature. FA can efficiently solve continuous as well as discrete combinatorial optimization problems. The firefly algorithm is suitable for unimodal as well as multimodal problems. It is found to be more efficient in dealing with multimodal optimization problems. Hybrid algorithms can be investigated, in which the firefly algorithm is combined with other optimization algorithms to improve the performance.

References

1. X.-S. Yang, *Engineering Optimization: An Introduction with Metaheuristic Applications*, Chichester: Wiley, 2010.
2. X.-S. Yang, Firefly algorithm, Lévy flights and global optimization, In: *Research and Development in Intelligent Systems*, XXVI, M. Bramer, R. Ellis, M. Petridis (eds). London: Springer, pp. 209–218, 2010.
3. J. Kwiecien, B. Filipowicz, Firefly algorithm in optimization of queueing systems, *Bulletin of the Polish Academy of Sciences*, Technical Sciences, Vol. 60, No. 2, 2012.
4. Ming-Huwi Horng, Vector quantization using the firefly algorithm for image compression, *Expert Systems with Applications*, Vol. 39, pp. 1078–1091, 2012.
5. R. B. Francisco, M. F. P. Costa, A. M. A. C. Rocha, Experiments with firefly algorithm, In: *Computational Science and Its Applications, ICCSA 2014, Lecture Notes in Computer Science*, Vol. 8580, B. Murgante et al. (eds). Cham: Springer, 2014.
6. X.-S. Yang, Xingshe He, Firefly algorithm: Recent advances and applications, *International Journal of Swarm Intelligence*, Vol. 1, No. 1, pp. 36–50, 2013.
7. X.-S. Yang, *Firefly Algorithms for Multimodal Optimization*, LNCS 5792, pp. 169–178, Berlin, Heidelberg: Springer-Verlag, 2009.

13

Bat Algorithm

13.1 Introduction

The nature-inspired optimization algorithms are powerful techniques for solving NP-hard problems that are found to be intractable for traditional algorithms. They are based on the study of natural phenomena such as biological evolution, and physical and chemical processes. Almost all of these algorithms use metaheuristics so that a solution close to the optimum can be attained in reasonably finite time. If the algorithm is going to do an exhaustive search of possible solutions to a problem the time and computational complexity will be high. In order to reduce the time and resources consumed some amount of heuristics becomes necessary. Nature inspires us with amazing techniques to provide solutions to complex engineering design problems in the areas of civil and mechanical engineering, electronics, communication, computer science, economics, management, and other related areas. The possibilities of adopting natural phenomena in our problem-solving techniques are enormous, and lots of research has been undertaken in this field ever since the genetic algorithm was invented and several decades later the particle swarm optimization algorithm was developed.

There are millions of species of flora and fauna in nature with each one having its own system of foraging and survival in the hostile environment. Study of these biological systems is fascinating, and several algorithms have been proposed based on these processes and characteristics. It is amazing how these simple creatures solve problems and survive by following simple rules and complex social interactions within their group. Each species has its own method of searching for food in the environment and communicating with their fellow members. They also share the food, mate with each other, and guard themselves and their group members against predators. Every animal, bird, or insect species follows some simple rules either individually or as a group, typically called a swarm. These biological entities exhibit swarm behavior, and the entire swarm is disciplined and undertakes activities for the benefit of the members of the whole group.

Bats are fascinating mammals that have wings to fly. They are nocturnal animals, and they also exhibit social interactions amongst themselves. They have a peculiar echolocation capability that is unique to their species. The bats use this echolocation principle to find their food which could be fruits or insects. Since they use the Doppler effect to detect and locate their prey, they must definitely have a signal-processing capability that is different from other animals and insects. The bats navigate themselves without bumping into obstacles during any part of the day. This capability of bats has led to the development of the bat algorithm (BA) which is one of the superior nature-inspired optimization techniques. The characteristics and behavior of bats in nature and the bat optimization algorithm [1] have been discussed in the following sections.

13.2 Behavior of Bats in Nature

Bats are mammals (order *Chiroptera*) that are capable of sustained flight. Their forelimbs are adapted as wings and their spread-out digits are covered with a thin membrane that helps them fly. There are more than 1200 species found throughout the world that constitute around 20% of mammal population. The largest bats are the flying foxes that weigh around 1.5 kg with a wingspan of about 2 m whereas the smallest are the bumblebee bats with a weight of around 1.5 g and the hog-nosed bats with a wingspan of 15 cm. Figure 13.1 shows the Indian flying foxes in Madhya Pradesh and Figure 13.2 shows spectacled flying foxes – male, female, and their young one – hanging upside down from a tree. Bats are nocturnal animals and mostly live in caves. They are traditionally divided into two categories, fruit-eating megabats and echolocating microbats, but they are also categorized based on other characteristics. They mainly feed on insects while some of them feed on fruits.

The bats feed on fruits and insects, sometimes even animals, with the vampire bats feeding on blood. They are important pollinators for plants especially in the tropical region. Pollination and feeding on insects are two important advantages of bats to humans. Bats largely feed on insects and balance the population as well as consuming the pests. A few of them feed on fruits like bananas and figs. But the disadvantage is that bats are carriers of rabies and other viruses. Bats mostly roost during daytime and forage during night time. Bats are blind, but they sense their environment through ultrasonic waves. Microbats and a few megabats use echolocation to sense the environment in the dark. Microbats use echolocation (sonar) extensively to sense the environment, avoid obstacles, detect prey, and find their roosting crevices. When bats are not flying, they hang upside

FIGURE 13.1
Indian flying foxes (*Pteropus giganteus*), Madhya Pradesh, India. (Author: Charles J, Sharp – own work, CC BY-SA 4.0. https://creativecommons.org/licenses/by-sa/4.0/deed.en.)

FIGURE 13.2
Spectacled flying foxes. (Author: Justin Welbergen, CC BY-SA 3.0. https://creativecommons.org/licenses/by-sa/3.0/deed.en.)

down from trees with their feet – a posture called *roosting*. Megabats roost with their head tucked towards the belly whereas microbats roost with their neck curled towards their back. Figure 13.3 shows an intermediate roundleaf bat (*Hipposideros larvatus*) roosting flock in a cave, in Lampung, Southern Sumatra.

  Bats transmit a pulse of ultrasonic frequency, and based on the received signal or pulse they build a map of their surroundings. The typical duration of the pulse ranges from 5 to 20 ms. The bats normally emit 10 to 20 bursts every second, and this can go up to 200 bursts when they are nearing their prey. The velocity of sound is 340 m/s and the wavelength of the pulses emitted is $\lambda = v/f$ (meters).Therefore the wavelength range of the emitted pulses is 3 to 14 mm. This is usually the typical size of the prey of bats. The brain and auditory system is responsible for comparing the transmitted and echo pulses and producing detailed images of the environment within the vicinity of their surroundings. This makes possible the detection and location of prey in darkness. Figure 13.4 shows ultrasound pulses emitted by a bat and the echo received from nearby objects through the left and right ear of the bat. Figure 13.5 also demonstrates echolocation in bats where A is the bat, B is the prey, d is the distance between bat and prey, E is the emitted wave of bat, and R is the received echo from the prey.

  Bats are one of the animals that make the loudest sound, ranging from 60 to 140 dB. The sound is loudest when searching for prey, and it reduces when the bat is approaching the prey. The frequency of emission of microbats is in the range of 25 to 100 kHz, and might go up to 150 kHz, extending beyond the range of human hearing. The range of travel of the emitted waves is a few meters. Microbats have the capability to detect obstacles that are as thin as human hair. Figure 13.6 shows the picture of a little brown bat flying during the day.

  Some bat species have fleshy extensions around the nose called nose-leaves that play a role in sound transmission. In low-duty cycle echolocation, bats make their transmission time short so that their transmission and echoes can be separated in time. These bats use a constant

FIGURE 13.3
Bats roosting in a cave. (Author: Wibowo Djatmiko – own work, CC BY-SA 3.0/2.5/2.0/1.0. https://commons
.wikimedia.org/wiki/Commons:GNU_Free_Documentation_License,_version_1.2. https://creativecommons.o
rg/licenses/by-sa/3.0/deed.en.)

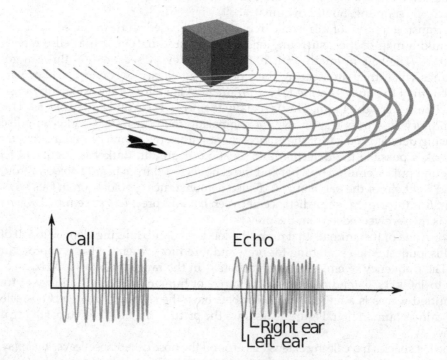

FIGURE 13.4
Ultrasound signals emitted by bat and echo received. [Author: Petteri Aimonen – own work (Public Domain).]

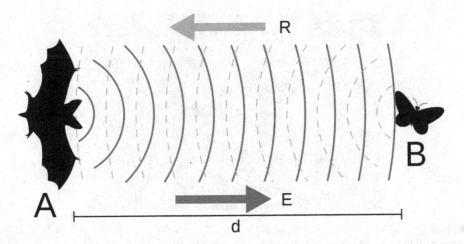

FIGURE 13.5

Echolocation in bats. (From Chiroptera echolocation: Augusto, Shung, Marek. Available at: https://commons. wikimedia.org/wiki/File:Chiroptera_echolocation.svg. Creative commons: https://creativecommons.org/ licenses/by-sa/3.0/deed.en)

frequency for transmission. Some of them use pulses whose frequencies span one octave. The bandwidth of the transmitted signal depends on the species. They contract their middle ear muscles during transmission so that they are not deafened by the sound. The muscles are relaxed after transmission so that the echoes can be heard when they return. The time delay between the transmitted and received pulse is used in calculating the range of the location of prey. In high-duty cycle echolocation, the transmission frequencies are lower and hence the transmission time is longer. The transmitted and returning echoes are separated in frequency. The bats are tuned sharply to the returning frequency, whereas they cannot hear the transmitted frequency and hence avoid deafening themselves. The Doppler shift of the returning echo allows the bats to estimate the movement, range, and location of the prey. When the bats are in flight they adapt themselves to change the frequency so that the echoes can still be detected and heard. Elevation of the target is estimated from the interference pattern of the received pulses. They also have the ability to passively listen to sounds made by insect movements on the ground or the flying of moths and other insects. Bats use their vision for traveling. Bats can differentiate between different types of insects in complete darkness using echolocation. Echolocation is effective only for short distances. Scanning by ultrasound waves is done repeatedly by the bats to construct a map of the environment. Bats build a 3D model of their environment using the time difference between emitted and received pulse, using their own two ears and the loudness of the echo. The Doppler shift caused by moving prey (such as wing flutter of insects) allows bats to detect the speed of movement, distance, orientation, type, and size of the prey. Some bats also have a good sense of smell.

Microbats have small eyes with mesopic vision that allow detection at low light levels. Mesopic vision is a combination of photopic and scotopic vision. Megabats have good eyesight with photopic vision that can also detect color in good lighting conditions and is as good as humans'. Microbats have large ears with a tragus that is important for echolocation, whereas megabats have comparatively smaller ears with no tragus. Bats are sensitive to earth's magnetic field and avoid flying in the sun to prevent overheating. They mostly roost during the hottest part of the day. Some bats are solitary, but most of them live in colonies of hundreds of thousands. The bat optimization algorithm has been developed

FIGURE 13.6
Little brown bat in flight. (Author: Andy Reago and Chrissy McClarren, CC BY 2.0. https://creativecommons.org/licenses/by/2.0/deed.en.)

based on the study of the echolocation characteristics and behavior of bats in nature that is described in the following sections.

13.3 Bat Optimization Algorithm

The bat algorithm (BA) is a metaheuristic algorithm based on the echolocation behavior of bats. In developing the algorithm the following simplifying assumptions are made:

- Bats fly randomly with a fixed velocity and take up different positions. Their pulse emissions have varying frequencies, wavelengths, and loudness which they use to search for prey. The rate of pulse emissions and the frequency are adjusted based on the distance of the bat to the prey.
- Bats use echolocation to detect as well as differentiate between prey (food) and other objects, and they are able to sense distance.
- The loudness of the ultrasonic pulses emitted by bats varies from a minimum value to a maximum value.

Let f_{min} and f_{max} be the minimum and maximum frequencies of the pulse emissions of the bat respectively. Correspondingly λ_{max} and λ_{min} are the maximum and minimum wavelengths of the bat emissions respectively. Using the formula, $\lambda = \dfrac{v}{f}$ the wavelength of the emission can be obtained from the frequency and *vice versa*, since the velocity v is a constant and is equal to 340 m/s. The bats normally emit in the higher frequency ranges, and high frequencies travel shorter distances of up to a few meters only. The rate of pulse emission could be assumed to be in the range [0, 1] where rate 0 represents no pulses emitted and 1 is the maximum rate of pulse emission.

13.3.1 Algorithm

Let the population size of the bats be N, the position of the ith bat in the d-dimensional search space be X_i^{iter} for the iteration indexed by $iter$, and the velocity of the bat at the current position and iteration is V_i^{iter}. The position of the bat in the d-dimensional search space is given by

$$X_i^{iter} = [x_{i1}^{iter} \ x_{i2}^{iter} \ .... \ x_{id}^{iter}] \tag{13.1}$$

The bats move in the search space searching for the global optimum which is nothing but the best solution to the problem. The equations governing the movement of the bats for exploration of the search space are:

$$X_i^{iter+1} = X_i^{iter} + V_i^{iter} \tag{13.2}$$

$$V_i^{iter+1} = V_i^{iter} + (X_i^{iter} - X_{gb}^{iter})f_i \tag{13.3}$$

where X_{gb}^{iter} is the global best position in the current iteration and f_i is the frequency of emission of the ith bat given by

$$f_i = f_{min} + r(i)(f_{max} - f_{min}) \tag{13.4}$$

where $r(i)$ is a random number drawn from the uniform distribution in the range $[0, 1]$, f_{min} is the minimum frequency of emission, and f_{max} is the maximum frequency of emission, and the limits on the frequency can be fixed according to the problem. During initialization, the frequencies are randomly assigned to the bats within the range chosen. Velocity is the product of wavelength and frequency, and either of them can be adjusted to update the velocity.

Every bat position is updated for local search in each iteration. The equation governing this exploitation movement of bats is given by

$$X_i^{iter+1} = X_i^{iter} + c.L_{av}^{iter} \tag{13.5}$$

where X_i^{iter+1} is the new position of bat i, X_i^{iter} is the old position of bat i, c is a random number that takes values in the interval $[-1, 1]$, and L_{av}^{iter} is the average loudness of all the bats in the population. The loudness and rate at which the pulses are emitted by the bats determine the intensity of the local search. The BA is similar to PSO in the position and velocity update equations. The loudness and rate of pulse emission of the bats vary with the distance of the bat to the prey during the search. As the bat approaches the prey the loudness of emission reduces and the rate of pulse emission increases. Let the loudness of bat i be given by L_i and the rate of emission of bat i be given by E_i. The minimum and maximum values for the loudness could be chosen by the user, where the minimum could even be zero if the prey is trapped, or it could be some other smaller number. The maximum value for loudness could be chosen to be any convenient number since it is going to be decreased with the increasing number of iterations. Typically the maximum value for L_i is taken as 100, but it could be less than that.

$$L_i^{iter+1} = \lambda . L_i^{iter} \tag{13.6}$$

where λ is a random number between 0 and 1. As $iter \to \infty$, $L_i^{iter} \to 0$. The rate of emission of the bat i is given by,

$$E_i^{iter+1} = E_i^0(1 - e^{-\eta(iter)}) \tag{13.7}$$

where η is a positive constant, and $E_i^{iter} \to E_i^0$ as $iter \to \infty$. If the new solution is better than the previous one, the loudness and emission rate are updated; this implies that the algorithm is approaching the optimal solution or the bat is approaching the prey.

13.3.2 Pseudocode

Initialization

   Choose the size of the bat population N

   Dimensions of the search space d

   Randomly initialize the position of the bats $X_i^{iter} = [x_{i1}^{iter}, x_{i2}^{iter}, ..., x_{id}^{iter}]$, $i = 1, 2, ..., N$

   Define Objective function $f(X)$

   Pulse frequency f_i, rate of pulse emissions E_i, and loudness of the emissions L_i

   Maximum number of iterations *MaxIter*

   iter = 1

while ($iter \leq MaxIter$) **do**

   Update the positions, velocities, and frequencies of the bats in the search space

   if ($r(i) > E_i$) **then**

      Select one solution among the best as determined by the fitness values

      Generate a local solution in the vicinity of the selected solution

   end if

   Generate a new solution by random movement (flying) of a bat

   if [($r(i) < L_i$) **and** ($f(X_i) < f(X_{gb})$)] **then**

      Accept the new solution

      Increase the rate of emission and reduce the loudness of emission

   end if

   Rank the bats according to their fitness values and find the current global best X_{gb}

   iter = *iter* + 1

end while

Bat with the highest fitness value is the global optimum solution

Flowchart

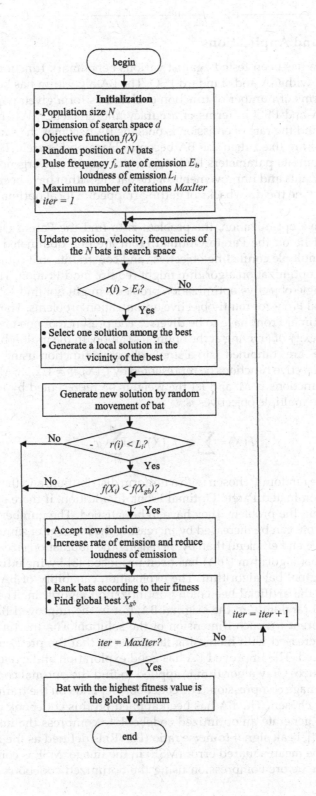

13.4 Variants and Applications

The bat algorithm has been tested against various benchmark functions and its performance compared with GA and standard PSO. The BA algorithm has been found to perform better in terms of number of function evaluations for a given population size. BA
is superior to GA and PSO in terms of accuracy and efficiency. When the loudness of
emission is zero and the rate of emission is one, the BA becomes the same as PSO. When
loudness and rate are fixed then the BA becomes Harmony Search. The proper adjustment of the appropriate parameters leads to a faster rate of convergence of the BA algorithm. Several variants and improvements of the bat algorithm have been proposed in the
literature to overcome the drawbacks of getting trapped in local optima and enhance the
performance.

In multi-objective optimization, the problem is to find the Pareto Optimal Front. All
the solutions that lie on the Pareto Optimal Front are non-dominated solutions. When
the problem has multiple conflicting objectives and constraints that could be either linear
or non-linear, the optimization algorithm might require modification. The algorithm that
works well for single-objective optimization problems might not find solutions that lie on
the Pareto Optimal Front for multi-objective optimization problems. The solutions that lie
on the Pareto Optimal Front have to be diverse. The presence of one or more constraints
adds to the complexity of arriving at the optimal solution. In the multi-objective BA [2], the
multiple objectives are combined into a single-objective function using a weighted sum.
Let the multiple objective functions be represented by $f_m(X)$, $m = 1, 2, ..., M$, where the number of objective functions is M, and let the weights be represented by w_m. The weighted
combination of the multiple objectives is:

$$F(X) = \sum_{m=1}^{M} w_m f_m(X) \text{ and } \sum_{m=1}^{M} w_m = 1 \qquad (13.8)$$

The weights can be randomly chosen with a uniform distribution, and they can be adjusted
to obtain a non-dominated Pareto Optimal Front. In addition, if there are linear or non-
linear constraints in the problem, they have to be satisfied. The number of points on the
Pareto Optimal Front can be increased by increasing the number of simulations (which is
more time-effective and efficient) than by increasing the population size.

The improved bat algorithm (IBA) has been proposed [3] by including three modifications to the original bat algorithm. The exploration capability of BA is improved by
hybridizing with the artificial bee colony optimization algorithm. The pulse frequencies are generated randomly in the original BA, but in the improved BA, the pulse frequencies are different for each dimension of the solution. The inertia weight factor is
also gradually decreased with increasing iterations so that the previous velocity effect
is gradually reduced. The improved BA has good exploration and exploitation capabilities. The Linde–Buzo–Gray algorithm is applied to find the optimal codebook in vector
quantization for image compression. The codebook depends on the training data set and
the initial vectors chosen. The BA has been applied on this codebook produced by the
LBG algorithm to generate an optimized codebook to compress the images, leading to
increased PSNR [4]. Peak signal-to-noise ratio (PSNR) is defined as the ratio of the peak
signal power to the mean squared error (MSE) in the image. MSE is computed between
the original image before compression using the optimized codebook and the decom-
pressed image.

The discrete version of the BA [5] has been proposed and applied to find the solution for the symmetric and asymmetric traveling salesman problem. BA can converge faster by switching between exploration and exploitation even in the early stages. But switching too quickly can also cause the algorithm to stagnate. Several variants of the BA have been proposed, and some of them are enumerated below [6]:

- The fuzzy logic bat algorithm introduces fuzzy logic into the BA.
- Multi-objective BA is an extension of BA for dealing with multi-objective problems.
- The *k*-means bat algorithm is a combination of *k*-means algorithm and BA for clustering applications.
- The chaotic bat algorithm uses Levy flights and chaotic maps to do parameter estimation.
- The binary bat algorithm is a binary version of the BA used for feature selection and classification.
- The differential operator and Levy flights bat algorithm has been proposed to solve function optimization problems.
- The improved bat algorithm is an extension of BA with Levy flights, and varying the parameters has given good results.

In addition to this, other hybrid versions of the BA have also been proposed for different applications. BA can deal with linear as well as non-linear continuous optimization problems efficiently. Combinatorial optimization problems which are generally considered to be NP-hard can also be solved using BA. It has also been found to be suitable for inverse problems. The combination of *k*-means and BA for clustering applications has been found to be superior to either of them considered separately.

BA has been applied for crop classification based on multispectral satellite images [7]. A clustering technique has been used in extracting information from the training samples and forming cluster centers. Crop classification is a challenging task since it involves lot of factors like geographical variation, weather conditions, crop yield, stage of growth, etc. BA has outperformed GA, PSO, and *k*-means clustering in crop classification performance. Topology optimization is finding the best geometric shape for a particular application or to meet a certain objective. When buildings are constructed, design of the optimal shape for structures, so that they can withstand the load with the least usage of construction materials and minimal cost, is an optimization problem. In microelectronics, the shape of the devices and their placement on the circuits or electronic boards determine the heat transfer which is an important issue in electronics. This makes the layout of devices on the board an optimization problem. BA has been found to be effective in solving the heat transfer problem by topology optimization [8].

Image segmentation is an important step in image processing applications. Segmenting the image into different regions is a challenging task since it involves fixing the optimum multilevel threshold for the segmentation. An objective function based on image entropy is formulated and the BA is applied to find the optimum multilevel thresholds [9]. Metaheuristic BA finds the optimum threshold with less time complexity compared to other classical and heuristic algorithms. In [10] a differential operator is introduced to increase the rate of convergence of BA. The Levy flight trajectory ensures that the algorithm is able to jump out of local optima. In problems with higher dimensions, this new strategy is more effective than the original BA.

13.5 Summary

BA is more powerful than GA and PSO because it includes the important characteristics and advantages of these algorithms as well as of other nature-inspired algorithms such as Harmony Search and Simulated Annealing. The echolocation characteristic of bats is very innovative compared to other insects, birds, and animals and proves that the bats have good signal-processing power. Utilizing this capability of bats in the optimization algorithm elevates the performance of the algorithm when applied to complex problems. The associated parameters of frequency of pulses emitted, rate of emission, and loudness of emission, can be adjusted and fine-tuned to get varying performances. The right combination of these parameters plays a key role in finding the global optimum solution to the problem. The bat constructs a three-dimensional map of the surrounding environment (search space) from the emitted pulses and the echoes received. It utilizes the Doppler effect in building this map. This map helps the bat in locating prey and obstacles so that the bat can navigate virtually in the dark. By adjusting the average loudness and rate of emission the BA effectively reduces to either PSO or Harmony Search. A population size of 20 to 50 and 100 iterations are sufficient for most of the applications. If necessary the number of iterations can be increased or a termination criterion can be included.

BA is simple, efficient, easy to implement, flexible, and can be applied to a wide range of problems. Some of the important applications include feature selection and classification, image processing, clustering, data mining, and job scheduling.It is efficient in finding the optimum solutions for NP-hard problems such as the TSP. It is similar to PSO and Harmony Search and requires fine-tuning of a few parameters only. When a bat is nearing its prey, it increases the rate of emission and reduces the loudness of emission. This capacity of bats in *zooming-in* on the prey is useful for exploitation of the search in the vicinity of the optimum and leads to early convergence. Moreover, in BA the parameters can be fine-tuned or modified adaptively as the bat is approaching the prey, that is, as the algorithm is nearing the optimal solution. As the iterations proceed the algorithm automatically changes from the exploration to exploitation mode. This makes BA more efficient than other algorithms.

References

1. X.-S. Yang, A new metaheuristic bat-inspired algorithm, In: *Nature Inspired Cooperative Strategies for Optimization (NISCO 2010)*, J. R. Gonzalez et al. (ed), Studies in Computational Intelligence, 284. Berlin: Springer, pp. 65–74, 2010.
2. X.-S. Yang, Bat algorithm for multi-objective optimization, *International Journal of Bio-Inspired Computation*, Vol. 3, No. 5, pp. 267–274, 2011.
3. Selim Yilmaz, Ecir U. Kucuksille, Improved bat algorithm on continuous optimization problems, *Lecture Notes on Software Engineering*, Vol. 1, No. 3, pp. 279–283, August 2013.
4. Chiranjeevi Karri, Umaranjan Jena, Fast vector quantization using a bat algorithm for image compression, *Engineering Science and Technology: An International Journal*, Vol. 19, pp. 769–781, 2016.
5. Eneko Osaba, Xin-She Yang, Fernando Diaz, Pedro Lopez-Garcia, Roberto Carballedo, An improved discrete bat algorithm for symmetric and asymmetric traveling salesman problems, *Engineering Applications of Artificial Intelligence*, Vol. 48, pp. 59–71, 2016.

6. X.-S. Yang, Bat algorithm: Literature review and applications, *International Journal of Bio-Inspired Computation*, Vol. 5, No. 3, pp. 141–149, 2013.
7. J. Senthilnath, Sushant Kulkarni, J. A. Benediktsson, X.-S. Yang, A novel approach for multi-spectral satellite image classification based on the bat algorithm, *IEEE Geoscience and Remote Sensing Letters*, Vol. 13, No. 4, pp. 599–603, 2016.
8. X.-S. Yang, Mehmet Karamanoglu, Simon Fong, Bat algorithm for topology optimization in microelectronic applications, *IEEE First International Conference on Future Generation Technologies*, pp. 150–155, 2012.
9. Adis Alihodzic, Milan Tuba, Bat algorithm for image thresholding, *Recent Researches in Telecommunications, Informatics, Electronics and Signal Processing, IEEE 12th International Conference on Signal Processing*, pp. 364–369, 2013.
10. Jian Xie, Yongquan Zhou, Huan Chen, A novel bat algorithm based on differential operator and Levy flights trajectory, *Computational Intelligence and Neuroscience*, Vol. 2013, pp. 1–13, 2013.

14

Flower Pollination Algorithm

14.1 Introduction

Nature-inspired metaheuristic optimization algorithms have been developed by mimicking the evolutionary processes in nature that occur among plants, animals, birds, insects, and other biological organisms. One such optimization technique is the flower pollination algorithm (FPA) based on the pollination process of flowering plants [1]. Pollination is a reproductive strategy in plants that transfers pollen from the male part of the flower to the female part of the same or a different flower. These pollen gametes reach the ovary where they are fertilized and develop into seeds. The seeds germinate and grow into new plants. The pollinating agents are wind, water, insects, birds, and animals, and also could be the plants themselves. The in-depth study of the pollination process and its types has led to the development of the flower pollination optimization algorithm which is competitive in performance with existing metaheuristic algorithms. FPA is a population-based metaheuristic algorithm that searches for the optimum solution in the search space in parallel with multiple agents.

The pollen-carrying agents might travel quite a distance before depositing their pollen in another flower. The distance traveled by the pollinating agents determines whether it is local or global pollination. There needs to be a balance between intensification and diversification in any optimization algorithm, and this balance is achieved by the movement of the agents carrying pollen. The movement of insects and birds can be modeled using Levy flights following a Levy distribution. The Levy flight is a path with straight lines punctuated by sharp 90° turns. The pollen can be deposited in the same flower or in different flowers of the same plant. The pollen can also be deposited in flowers of another plant belonging to the same species or a different species. The flower where the pollen gets deposited determines the types of pollination.

These optimization algorithms that are based on the evolutionary process in nature are able to solve complex non-linear engineering design problems that might be constrained. The algorithms are efficient and provide solutions that are quite close to the global optimum in finite time. The accuracy of the solution can be increased by increasing the number of iterations as almost all of these algorithms are iterative. The parameters involved have to be chosen properly so that the algorithm reaches the global optimum without getting stuck in local optimum. If the algorithm gets trapped in a local optimum it has to jump out of it and converge to the global optimum in finite time. The convergence rate of the algorithm is important in practical applications. The algorithm has to be able to find a solution for single-objective as well as multi-objective optimization problems. The FPA discussed in the following sections also follows the Darwinian principle of *survival of the fittest* as the plants evolve based on the reproduction process in nature.

14.2 Flower Pollination

Nature has been evolving over millions of years, and biological systems are becoming more robust and efficient. One of the important processes that has existed and evolved over the years is reproduction. The reproduction process takes place among all living beings, including plants, animals, birds, insects, and humans. Depending on the species, there is a large variation in the reproduction mechanism. In plants, the process of reproduction is through pollination. The pollen from the anther (male portion) of a flower gets transferred to the stigma (female portion) of either the same or a different flower. This process is called fertilization. The fertilization process ultimately leads to the production of fruits and seeds by the plants. These seeds later grow into new plants and thus the species reproduces.

Pollination is of two types – self-pollination and cross-pollination. In self-pollination, the pollen is transferred to flowers of the same plant, whereas in cross-pollination, the pollen is transferred to flowers of a different plant. The transfer takes place by means of agents like wind, birds, bees, and other insects and animals. The insects like butterflies or bees that visit the flower to take nectar or simply sit on the flower get dusted in pollen or, in other words, the pollen gets stuck to the insect. When the insect visits another flower either on the same plant (self-pollination) or on another plant (cross-pollination) the pollen is deposited on the stigma of the flower. Figure 14.1 shows a hibiscus flower with distinct anther and stigma, and a spring rose flower is shown in Figure 14.2.

FIGURE 14.1
Hibiscus flower (*Hibiscus Fragilis*) at Kew Gardens, London. (Author: C. T. Johansson – own work, CC BY-SA 3.0. https://creativecommons.org/licenses/by-sa/3.0/deed.en.)

FIGURE 14.2
Spring rose (*Helleborus Orientalis*). (Author: Dominicus Johannes Bergsma – own work, CC BY-SA 3.0. https://cr eativecommons.org/licenses/by-sa/3.0/deed.en.)

When pollen is deposited on the stigma it goes down the style (tube) until it reaches the ovaries, and fertilization takes place in the ovaries of the flower. This leads to the development of seeds and propagation of the species. Sometimes the seeds are embedded within the fruit of the plant, but they contribute to reproduction. Figure 14.3 shows the male and female parts of a mature flower. The reproductive parts of the flower such as the style,

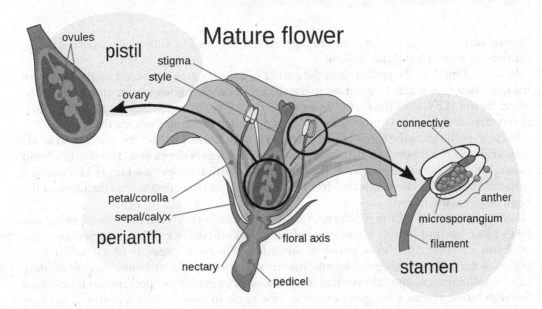

FIGURE 14.3
Illustration of the parts of a mature flower. [Author: Mariana Ruiz LadyofHats – own work (Public Domain).]

FIGURE 14.4
Christmas lillium. 1. Stigma, 2. style, 3. anthers, 4. filament, 5. sepal. (Author: J. J. Harrison – own work, CC BY-SA 3.0. https://creativecommons.org/licenses/by-sa/3.0/deed.en.)

stigma, anther, pollen, and ovaries are highlighted. Figure 14.4 shows the different reproductive parts of a Christmas lillium.

In self-pollination, the pollen from the anther of a flower gets deposited on the stigma of the same flower or it gets transferred to the stigma of another flower belonging to the same plant. Figure 14.5 shows the orchid flower (*Ophrys apifera*) in which automatic self-pollination occurs. It shows two of the anthers carrying pollen bending towards the stigma.

In cross-pollination the birds or insects that visit the flowers collect the pollen and when they visit another flower of a different plant, the pollen gets deposited. The flowers could belong either to the same species or they might be of different species. Figure 14.6 shows a carpenter bee with pollen collected from the flower night-blooming cereus (princess of the night or Honolulu queen).

Pollination mostly occurs within species, but pollination between different species also takes place leading to hybrid varieties of flowers. Pollination can be classified as biotic or abiotic. In abiotic methods, pollen is carried by wind and water. In biotic pollination, pollen is carried by biological agents like insects, birds, and animals. Approximately 10% of pollination is done through abiotic means whereas 90% of pollination takes place through biotic agents. One good example of a plant in which abiotic pollination takes place is grass. Biotic pollinators are living organisms that carry pollen from one flower to another. There are more than two lakh species of organisms that carry out biotic

FIGURE 14.5
Orchid flower (*Ophrys apifera*). [Author: Bernard DUPONT from France. Source: Bee Orchid (*Ophrys apifera*), CC BY-SA 2.0. https://creativecommons.org/licenses/by-sa/2.0/deed.en.]

pollination. They include butterflies, bees, and bats to name a few. The pollinators like honey bees visit only certain species of flowers and pollinate amongst them. This leads to reproduction and propagation of the same species of flowers leading to what is called *flower constancy* [2]. Such insects visit only certain flower patches, and in the process, bypass valuable food sources. One possible reason for such behaviour is the memory of insects and their ability to remember characteristics of certain species of flowers only. Insects which visit several species of flowers either have good enough memory to remember all of them or do not have memory to remember any of them. The distance of pollination depends on the pollinating agent and the distance it can travel, especially through flying. Figure 14.7 shows a honey bee collecting nectar from a flower and during the process, pollen gets stuck to its rear leg that is clearly visible in the picture. Tiny hairs on the body of the bee have a slight electrostatic charge, causing pollen from the flower to stick to the bee's body which could get deposited on another flower when the bee visits that flower. This is in aid of cross-pollination in which the bee could visit flowers either from the same species or different species. Figure 14.8 shows a hummingbird that typically feeds on the red flower, leading to *flower constancy*. The advantage of *flower constancy* is maintaining the integrity of the species.

There are millions of plants, and more than 75% of them are flowering plants. Flowers are associated with the reproduction of the plant species through pollination, and they later produce seeds along with fruits. The pollination process has evolved with some plants or flowers being very attractive or producing nectar or scents to attract certain types of insects. Some of the flowers have specialized color or odor to trap insects or birds. They also have traps to lure and capture unwary insects. They have been found to be very effective in luring insects for pollination.

The combination of the above factors leads to optimal reproduction of the flowering plants that ultimately leads to survival of the fittest.

FIGURE 14.6
Carpenter bee with pollen collected from night-blooming cereus. (Author: Brocken Inaglory – own work, CC BY-SA 3.0, 2.5, 2.0, 1.0. https://commons.wikimedia.org/wiki/Commons:GNU_Free_Documentation_License,_version_1.2. https://creativecommons.org/licenses/by-sa/3.0/deed.en.)

FIGURE 14.7
Honey bee on flower with pollen collected on rear leg. (Author: Michael Palmer – own work, CC BY-SA 4.0. https://creativecommons.org/licenses/by-sa/4.0/deed.en.)

FIGURE 14.8
Hummingbird feeding on the red flower. (Author: Brocken Inaglory – own work, CC BY-SA 3.0/2.5/2.0/1.0. https://commons.wikimedia.org/wiki/Commons:GNU_Free_Documentation_License,_version_1.2. https://creativecommons.org/licenses/by-sa/3.0/deed.en.)

Figure 14.9a shows a young flower of the species *Geranum incanum* with its anthers ready to open and its pistil (stigma, style, ovary) not yet developed. When the flower opens its anthers, it changes color to attract pollinators, and when it matures fully it sheds its anthers and stamen in order to avoid self-pollination. Figure 14.9b shows a group of mature as well as young flowers of the species *Geranum incanum*. The flower at the top is mature with its anthers shed whereas the lower ones still have anthers. The stigma is not yet fully developed, and there is a change in the color of the flowers that indicates to the pollinators that it is ready to receive pollen. Figure 14.9c shows the fully mature flower of *Geranum incanum* with its anther and stamens shed and stigma deployed to receive foreign pollen. Some of the flowering plants have a specialized mechanism to attract certain species of birds or insects that aid in pollination and hence propagation of the same species.

14.3 Flower Pollination Optimization

An optimization algorithm based on the flower pollination process has been proposed by Xin-She Yang in 2012 called the *flower pollination algorithm* (FPA). The process and characteristics of flower pollination are idealized into four basic rules upon which the algorithm is developed:

- In biotic and cross-pollination the pollen-carrying agents are assumed to take Levy flights and hence this is global pollination.

- In abiotic and self-pollination the pollen is carried over a small distance; hence it is categorized as local pollination.

(a)   (b)

(c)

FIGURE 14.9
(a) Young flower of *Geranum incanum* species with anthers and no pistil. (b) *Geranum incanum* flowers – top flower is mature and lower ones are young. (c) Mature flower of *Geranum incanum* ready to receive foreign pollen. (Author: Jon Richfield – own work, CC BY-SA 3.0. https://creativecommons.org/licenses/by-sa/3.0/deed.en.)

- In *flower constancy* reproduction is among the same species of flowers. This is included in the algorithm as reproduction probability which is proportional to the similarity of the two flowers (source and receiver of pollen).

- The proportion of local and global pollination is determined by a parameter p_s that assumes values in the interval [0, 1].

Local pollination takes up a significant proportion of the total pollination activities, since wind and water are two of the pollen-carrying agents. Moreover, the flowers of the same plant are in close proximity, and this factor aids in local pollination. Birds, insects, bees, flies, and other biotic agents can fly for a long distance, and their movement can be modeled by Levy flight behavior with the Levy distribution. Every plant has numerous flowers, and each flower has hundreds or thousands of pollen gametes. To simplify the development of the optimization algorithm, it is assumed that every plant has one flower and every flower has one pollen gamete. This makes the plant, flower, and pollen gamete equivalent to each other, and they represent one possible solution to the problem in the case of single-objective optimization. This concept can be extended to include multiple flowers and multiple pollen gametes, to solve multi-objective optimization problems.

14.3.1 Algorithm

Let X_i^{iter} be a vector representing a pollen gamete or a flower/plant in the d-dimensional search space during the iteration *iter*. The population size is N.

$$X_i^{iter} = [x_{i1}^{iter} \ x_{i2}^{iter} \ .... \ x_{id}^{iter}] \ i = 1, 2, ..., N \tag{14.1}$$

The algorithm is assumed to be iterative, indexed by the variable *iter*, and the maximum number of iterations is *MaxIter*. The solution vector evolves in the next iteration as given by Equation 14.2 which represents global pollination.

$$X_i^{iter+1} = X_i^{iter} + L(u)(X_i^{iter} - X_{gb}^{iter}) \tag{14.2}$$

where $L(u)$ is the step size that follows Levy distribution and X_{gb} is the global best solution in the current iteration. This Equation 14.2 represents global pollination that is biotic, since Levy flights are undertaken by birds and insects that can fly long distances. $L(u)$ is the Levy distribution that models the transition probability, thus making the next position depend on the present position and the transition probability. The Levy distribution is approximated and given in Equation 14.3.

$$L(u) \approx t^{-u}, \ 1 \leq u \leq 3 \tag{14.3}$$

The second rule of the algorithm is given by Equation 14.4 that represents local pollination and *flower constancy*:

$$X_i^{iter+1} = X_i^{iter} + r(i)(X_j^{iter} - X_k^{iter}) \tag{14.4}$$

where X_j^{iter} and X_k^{iter} are pollens from different flowers of the same species leading to flower constancy, and $r(i)$ is a random number that takes on values in the interval [0, 1] with a uniform distribution. This parameter $r(i)$ models stochasticity in the algorithm and simulates a local random walk. Usually flower patches contain flowers of the same species that could be in the immediate vicinity or in the neighborhood of the flower represented by X_i. Sometimes the flowers belonging to the same species might be far away from the neighborhood of X_i. Let p_s be a probability that is equivalent to a switch whose value determines whether it is local or global pollination. Smaller values of p_s (<0.5) make it a local search, whereas if p_s is increased ($p_s > 0.5$), it becomes global pollination (global search). The value of $p_s = 0.8$ is optimum for most of the optimization problems.

The two key steps in the algorithm are based on global pollination and local pollination along with flower constancy and Levy flights. In the first step, pollen is carried by birds and insects that fly over long distances. This is biotic global pollination, and the global best solution is represented as X_{gb}. In the second rule the pollination is local along with *flower constancy* that is modeled by a local random walk. FPA has been found to perform better than GA and PSO comparatively in terms of the number of iterations required for attaining the global optimum solution.

14.3.2 Pseudocode

Initialization

  Population size N of flowers/pollen gametes

  Initial random position of the N flowers/pollen gametes in the search space

  Define objective function $f(X)$ that is d-dimensional,

   where $X_i = [x_{i1} \ x_{i2} \ ...... \ x_{id}]$, $i = 1, 2, ..., N$

  Evaluate the objective function for $i = 1, 2, ..., N$ and identify the global best X_{gb}

  Define switch probability p_s with an initial random value in the interval [0, 1]

  Define maximum number of iterations *MaxIter*

  iter = 1

while (*iter* ≤ *MaxIter*) **do**

 for i = 1 to N

 Generate random number r_s in the interval [0, 1]

 if ($r_s < p_s$) **then**

  Obtain a step size $L(u)$ from the Levy distribution

  Perform global pollination using Equation 14.2

 else if ($r_s \geq p_s$)

   Generate random number $r(i)$ in the interval [0, 1]

   Randomly choose X_j and X_k from the population

   Perform local pollination using Equation 14.4

 end if

 Evaluate new solutions and if better, update the population of pollen/flowers

 end for

 Evaluate the solutions to find the current best X_{gb}

 iter = *iter* + 1

end while

Current best solution X_{gb} is the global optimum solution

Flowchart

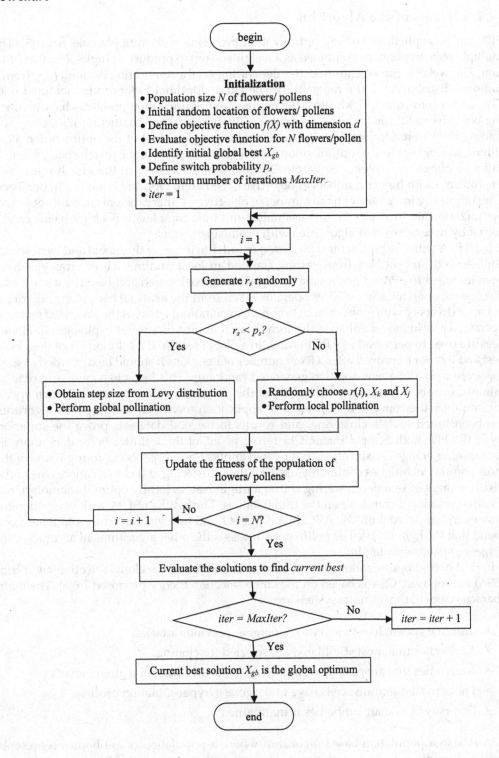

begin

Initialization
- Population size N of flowers/ pollens
- Initial random location of flowers/ pollens
- Define objective function $f(X)$ with dimension d
- Evaluate objective function for N flowers/pollen
- Identify initial global best X_{gb}
- Define switch probability p_s
- Maximum number of iterations *MaxIter*
- $iter = 1$

$i = 1$

Generate r_s randomly

$r_s < p_s$?

Yes

No

- Obtain step size from Levy distribution
- Perform global pollination

- Randomly choose $r(i)$, X_k and X_j
- Perform local pollination

Update the fitness of the population of flowers/ pollens

No

$i = i + 1$

$i = N$?

Yes

Evaluate the solutions to find *current best*

$iter = MaxIter$?

No

$iter = iter + 1$

Yes

Current best solution X_{gb} is the global optimum

end

14.4 Variants of the Algorithm

FPA can be applied for solving optimization problems with multiple objectives [3]. The multiple objectives can be combined as a weighted sum to produce a single-objective function. The weights can be chosen either depending on the application or randomly from a uniform distribution in the range [0, 1]. The Pareto Optimal Front can be obtained with FPA, and it converges quickly. Real-life engineering optimization problems have conflicting objectives with or without constraints. Combining these conflicting objectives and finding the Pareto Optimal Front is quite challenging for most of the optimization algorithms. In single-objective optimization, the solution is a point in the search space, whereas in a two-objective problem, the Pareto Optimal Front is a curve in the search space, and in problems with higher numbers of objectives, the Pareto Optimal Front is a hyperspace. The complexity increases with the number of objectives. Other problems in multi-objective optimization are increases in dimensionality and time complexity. Such problems can be solved by nature-inspired algorithms with promising results.

In [4] FPA with bee pollinator is proposed in order to improve the global and local search abilities and prevent FPA from getting trapped in local minima. Three strategies have been included in FPA for improving the local and global search abilities. The *discard solution (pollen)* operator and *crossover* operator taken from the artificial bee colony algorithm enhance diversity with global search (improve exploration) whereas the *elite-based mutation* operator is included to enhance the local search ability (improves exploitation). Honey bees are used to perform Levy flights and do a global search. If a solution is not the global best and it is not improved after a fixed number of iterations it should be discarded (*discard pollen operator*) and a new solution generated randomly. This helps in coming out of local minima. *Crossover* increases the diversity of the population. *The elite-based mutation* operator improves the convergence speed. The application for which the proposed algorithm has been tested is data clustering, and results for several data sets prove the superiority of the FPA with bee pollinator. Clustering is one of the techniques for data analysis, data mining, image classification, and related applications. *k-means* clustering is one of the most popular techniques commonly used for data clustering and classification. One of the disadvantages of *k-means* clustering is that it might lead to locally optimal solutions since the final solution depends upon the initial values. The results of FPA with bee pollinator have been compared with DE, ABC, FPA, CS, PSO, and *k-means* clustering, and it has been found that the hybrid FPA–bee pollinator surpasses the other algorithms in accuracy, convergence speed, and stability.

In [5] a hybrid optimization algorithm based on FPA and the clonal selection algorithm (CSA) is proposed. CSA is based on the clone selection theory proposed in [6]. The main characteristics of the immune system are:

- Immune system has memory to remember previous attacks.
- Antibodies that most stimulated are selected for cloning.
- Antibodies that are poorly stimulated or not at all stimulated are removed.
- The activated immune cells have undergone a hyper-mutation process.
- Diversity of human antibodies is maintained.

CSA is also a population-based algorithm where a population of antibodies represents potential solutions in the search space. Antibodies that are activated by a certain foreign body only proliferate. Cloning makes copies of antibodies that have high affinity so that antibodies with higher affinity have a higher probability of being cloned. The clones are better matched with antigens through mutation. The mutated antibodies are mixed with the current population, and they are ranked to choose the best memory cells. Finally, the lowest-affinity antibodies are replaced with randomly chosen population members in order to enhance the diversity.

In the FPA the high-affinity solutions are cloned in proportion to their affinity before applying local pollination. A step size scaling factor is introduced into the local pollination step. In order to avoid getting stuck in a local minima, the algorithm checks whether the global best has changed in the last 100 iterations; if not the entire population is replaced with newly generated random solutions retaining the global best solution. The combination of the good explorative properties of FPA (global search) and the good exploitative properties of CSA (local search) through high fitness antibodies has made the hybrid modified FPA more efficient than either of the two algorithms taken separately. The performance of the modified FPA has been compared with five of the famous existing algorithms – Simulated Annealing, genetic algorithm, FPA, bat algorithm, and firefly algorithm. The FPS–CSA algorithm has been found to outperform the existing metaheuristic algorithms.

Hybrid FPA with PSO has been proposed [7] to improve the accuracy of the search process and convergence speed of the algorithm. This algorithm gives better performance for constrained optimization problems. Initially, the PSO algorithm is applied in the search space to find the best solutions. The best solutions found by PSO become the initial points for FPA. The constraints in the optimization problem are taken into account by formulating an overall function combining the actual objective function with the constraints using appropriate weights. Now it becomes a weighted combination of multiple functions including the objective function and the constraints. The superiority of the hybrid PSO–FPA algorithm has been validated with a set of well-known test problems. In [8] the FPA algorithm performance has been evaluated for continuous optimization functions and its properties studied. The performance of FPA has been compared with PSO. In [9] the mutation operator has been combined with FPA to develop five new variants of the FPA. Amongst all of them, the adaptive Levy FPA has been found to give good results. The proposed variants of FPA have been tested on 17 benchmark functions and compared with the artificial bee colony, firefly, gray wolf, differential evolution, and bat algorithms. The results have been presented for population sizes of 40, 60, and 80 with fixed dimensions. In [10] FPA has been compared with the bat algorithm and tested on ten of the standard unimodal and multimodal benchmark functions. From the experimental results it is seen that the FPA outperforms the bat algorithm with respect to quality of solutions, consistency, and convergence characteristics.

In [11] the FPA has been applied to optimize the lifetime of wireless sensor networks by optimizing the power used by each node. The clustering of the nodes is mathematically modeled as a continuous function that is unconstrained. Moreover, the cluster nodes are associated with the appropriate cluster head in an optimal manner. The distance between the sensor node and the cluster head is defined as the fitness function. This minimizes the total distance as well as the total energy consumed by the nodes. FPA has been found to perform better in terms of wireless sensor network lifetime and stability as compared to the classical low-energy adaptive clustering hierarchy (LEACH) algorithm. In [12] FPA has been applied to minimize the total power loss in distributed generation systems. The objective function is to minimize the total power loss of the distribution system. The equality constraint is that the power loss plus the total power demand should equal the power of the distribution generation system. The inequality constraint is that the bus voltage should lie between a minimum and maximum limit. These constraints are applicable for all the buses. The algorithm has been tested on three different systems, and the performance was found to be satisfactory.

14.5 Summary

This chapter has given a lucid description of the flower pollination algorithm that is based on the interesting characteristics of the pollination of flowering plants. The algorithm can be employed for unconstrained as well as constrained optimization problems. The algorithm is simple with few parameters to tune, and it can be applied for any complex engineering design problem. The number of parameters in FPA is very low, and its performance is comparable to or surpasses other metaheuristic algorithms such as GA and PSO. This simplicity makes it very popular for solving NP-hard optimization problems. This can be extended to the discrete space to solve combinatorial optimization problems also. A population size of around 20 to 50 with 100 iterations is suitable for most applications. The FPA for single-objective optimization can be extended to multi-objective optimization with the non-dominated solutions lying on the Pareto Optimal Front.

As an extension to the above, two hybrid algorithms have been proposed in the literature – hybrid FPA–CSA and hybrid PSO–FPA – with promising results. The results of the hybrid algorithms show that the combination of more than one metaheuristic algorithm improves the performance with respect to accuracy of the results and speed of convergence when compared to the performance of single metaheuristic algorithms. The ability of insects and birds to explore the search space by traveling longer distances leads to diversification and global exploration of a vast search space. The concept of *flower constancy* and self-pollination ensures intense local search and hence leads to the intensification property of the algorithm. The right balance and interaction between these two principles makes the algorithm efficient and effectively solves intractable NP-hard problems. The simple concept of one pollen gamete and one flower on one plant can be extended to multiple pollen gametes and multiple flowers depending on the applications.

References

1. Xin-She Yang, Flower pollination algorithm for global optimization, In: *Unconventional Computation and Natural Computation 2012, Lecture Notes in Computer Science*, Vol. 7445, pp. 240–249, 2012.
2. L. Chittka, J. Thomson, Nickolas M. Waser, Flower constancy, insect psychology and plant evolution, *Naturwissenschaften*, Vol. 86, Springer-Verlag, pp. 361–377, 1999.
3. Xin-She Yang, Mehmet Karamanoglu, Xingshi He, Multi-objective flower algorithm for optimization, *Procedia Computer Science*, Vol. 18, pp. 861–868, 2013.
4. Rui Wang, Yongquan Zhou, Shilei Qiao, Kang Huang, Flower pollination algorithm with bee pollinator for cluster analysis, *Information Processing Letters*, Vol. 116, pp. 1–14, 2016.
5. A. Emad Nabil, Modified flower pollination algorithm for global optimization, *Expert Systems with Applications*, Vol. 57, pp. 192–203, 2016.

6. F. R. Fekety, The clonal selection theory of acquired immunity, *Yale Journal of Biology and Medicine*, Vol. 32, p. 480, 1960.
7. O. Abdel Raouf, M. Abdel-Baset, I. El-henawy, A new hybrid flower pollination algorithm for solving constrained global optimization problems, *International Journal of Applied Operational Research*, Vol. 4, No. 2, pp. 1–13, Spring 2014.
8. Szymon Lukasik, Piotr A. Kowalski, Study of flower pollination algorithm for continuous optimization, In: *Intelligent Systems*, Editors: P. Angelov, K. T. Atanassov, L. Doukovska, M. Hadijski, V. Jotsov, J. Kacprzyk, N. Kasabov, S. Sotirov, E. Szmidt, S. Zadrozny, Springer, pp. 451–459, 2015.
9. Rohit Salgotra, Urvinder Singh, Application of mutation operators to flower pollination algorithm, *Expert Systems with Applications*, Vol. 79, pp. 112–129, 2017.
10. Nazmus Sakib, Md. Wasi Ul Kabir, Md. Subbir Rahman, Mohammad Shafiul Alam, A comparative study of flower pollination algorithm and bat algorithm on continuous optimization problems, *International Journal of Applied Information Systems*, Vol. 7, No. 9, September 2014.
11. Marwa Sharawi, E. Emary, Imane Aly Saroit, Hesham El Mahdy, Flower pollination optimization algorithm for wireless sensor network lifetime global optimization, *International Journal of Soft Computing and Engineering*, Vol. 5, No. 3, pp. 54–59, July 2014.
12. P. Dinakara Prasad Reddy, V. C. Veera Reddy, T. Gowri Manohar, Application of flower pollination algorithm for optimal placement and sizing of distributed generation in distribution systems, *Journal of Electrical Systems and Information Technology*, Vol. 3, pp. 14–22, 2016.

15

Gray Wolf Optimization

15.1 Introduction

Gray wolf optimization (GWO) is a metaheuristic optimization algorithm that has been modeled on the behavior of gray wolves in nature. Metaheuristic algorithms are simple and have an in-built stochasticity to solve complex problems that have been found to be intractable for traditional algorithms. They are simple to implement and attain the optimum solution with less time and computational complexity than their classical counterparts. Metaheuristic algorithms are not specific to a problem, but they produce higher quality solutions for certain problems compared to others. They can be applied to a diverse set of problems ranging over areas of engineering, computer science, economics, business, financial modeling, etc. Metaheuristics combine local search as well as global search over the space of solutions, and their randomness enables them to jump out of stagnation. This ensures diversity of the search and hence increases the chances of finding the global optimum. When the nature-inspired optimization algorithms are hybridized with other evolutionary algorithms they produce better results. Additionally, extended and improved versions of these swarm intelligence algorithms have also been proposed by several researchers.

Population-based search algorithms such as the gray wolf optimization are more efficient since they are able to undertake the search for the optimum in parallel through multiple agents, thus reducing the time taken. Gray wolves have a hierarchy within their pack, and they are a disciplined bunch of animals. The wolves obey their leaders and undertake the hunting and attacking of prey for the benefit of the entire pack. GWO emulates the hierarchy of the wolf pack and their hunting techniques employed. The hunting, chasing, encircling, and attacking of prey by gray wolves are the fundamental operations employed in the GWO algorithm. These operations are mathematically modeled into the design of the algorithm. GWO has been applied on several benchmark data sets and engineering design applications and found to be comparable in performance to the existing algorithms of particle swarm optimization, gravitational search algorithm, differential evolution, and evolutionary strategies and programming. GWO has been found to be simple, flexible, derivative-free, and efficient in attaining the global optimum solution for complex problems. Several variants and improvements over the GWO have been proposed, and they have also been discussed in this chapter.

15.2 Gray Wolf Characteristics

The gray wolf belongs to the *Canidae* family, and the species is *Canis lupus*. It is a canine mostly found in North America and some regions of Europe-Asia. Wolves have color

FIGURE 15.1
European gray wolf, Prague Zoo. (Source: www.flickr.com/photos/kachnch/16364273038, CC BY 2.0. https://cr
eativecommons.org/licenses/by/2.0/deed.en.)

ranging from white to gray to black. Figure 15.1 shows a European gray wolf at the Prague
Zoo, Czech Republic.

The habitats of wolves are deserts, forests, mountains, and swamps. Wolves are very territorial in nature, and they mark the boundaries of their territories, and claim and defend
them. Wolves defend their territories aggressively and mark it with the scent of their urine.
Wolves have territories ranging from tens to hundreds of square miles. Territory size is
mostly influenced by pack size, neighboring wolf packs, human habitat, etc. Wolves use
dens when they have young pups. Dens are dug in well-drained soil, mostly near water in
natural structures like boulders and tree logs. Wolves eat other animals like rabbit, moose,
deer, salmon, livestock, and their carcass, and sometimes they also feed on vegetation.
When their hunt is successful the wolves have a feast whereas if they are not able to find a
prey, they starve until they are able to get a kill.

Wolves live in groups of seven to eight, called packs, consisting of a father, mother, and
children. They live together, hunt, communicate, protect their territory, and raise their
young. Wolves howl to communicate among their pack as well as to wolves of other packs,
probably to warn of impending danger. They also howl for other types communication
such as claiming of territory, warning intruders, or identifying wolves. The howl is unique
for a pack, and they howl more during the full moon. The howl starts with a single wolf,
and others join in. Wolves communicate not only by howl but also through growling, bark-
ing, whimpering, whining, snarling, and through the scent of their bodies using their good
sense of smell. Wolves can sense smell more than a mile away. Communication amongst
them is very essential for their survival, and most of the communication is through body
language. When a wolf is happy, it prances about with its front lowered, body (hump) and

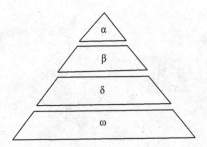

FIGURE 15.2
Social hierarchy of gray wolves.

tail raised. When a wolf is angry, it has a furrowed forehead, and might show its fangs and growl. Young wolves stay with their parents (in their pack) for at least two years before they start a new pack or join another pack. They have good body language to indicate whether they are dominant or subservient to another. If a wolf has its head and tail held high with perked up ears and baring teeth, it is confident and dominating. If a wolf has its tail between its legs, and is slinking towards another one with flattened ears, closed mouth, and slit-like eyes then it is approaching a dominant wolf. Wolves growl when they are angry and whine when they are affectionate.

Gray wolves live in packs, and they have a social hierarchy as indicated in Figure 15.2. There are four categories of wolves, namely alpha, beta, delta, and omega wolves, with alpha being the highest in the hierarchy followed by beta, delta, and omega respectively in a top-down manner. The leader of the pack is the alpha female wolf. The pack consists of one alpha female and one alpha male, and they mate and produce pups. The pups stay with the pack for two to three years and then move into nearby territory (~65 miles to 1000 miles) to form or join a new pack. Alpha wolves are the decision makers in the pack and decide when to hunt and move, and they always eat first when a prey is killed. Alphas are pack leaders that mark territory, locate and establish dens, hunt for prey, and lead the pack. Only alpha wolves are allowed to mate in the pack but that does not necessarily mean that they are the strongest wolves among the pack. Their dominance is acknowledged by the other wolves and the pack is disciplined. The beta wolf is at the next level to alpha. In case one of the alpha wolves passes away, a beta wolf in the pack will take over its position. It acts as the interface between the alpha wolf and other members of the pack. It is subservient to the alpha and reinforces the orders of the alpha to the other members of the pack. The feedback from the rest of the pack is collected by the beta wolf and passed on to the alpha wolf. The third level of wolves in the pack is deltas. They serve the functions of scouts, sentinels, hunters, and caretakers. Scouts watch the boundary of their territory; sentinels guard the pack against attacks. Omega wolves are the last in the hierarchy of the pack, and they have to be subservient to all the other members higher up in the hierarchy. Omega wolves are the babysitters in the pack, last to eat among all the wolves, and they contribute to maintaining the structure of the pack.

Wolves travel a lot and spend almost half their time traveling. Wolves help to maintain the diversity of the ecosystem, and their kills are also prey for other animals. The survival of the wolf population in the wild depends on the availability of prey, territorial disputes with other packs, human intervention, and other influences in the environment. The wolves normally hunt in packs or groups, and they exhibit social behavior. The characteristics of group hunting are: (i) tracking/chasing/approaching the prey, (ii) pursuing/encircling/harassing the prey, (iii) attacking the prey. Figure 15.3 shows a pack of wolves

hunting bison in Yellowstone National Park. Figure 15.4 shows wolves attacking prey (two wolves attack a moose at Isle Royale, Michigan, USA).

The gray wolf optimization algorithm is discussed in the following section.

15.3 Gray Wolf Optimization

The gray wolf optimization algorithm is based on the hierarchy and social interactions among the wolves belonging to a pack. The hunting techniques and disciplined behavior of gray wolves are the inspiration behind the gray wolf optimization algorithm. The activities of hunting, tracking, encircling, and attacking the prey are mathematically modeled in the optimization algorithm. The best solution is *alpha* (α), the second and third best solutions are *beta* (β) and *delta* (δ). The remaining solutions are the *omega* (ω). It is assumed that hunting is guided by α, β, δ wolves and the ω follows these three categories of wolves.

15.3.1 Gray Wolf Encircling Prey

Let X_i^{iter} represent the position vector of the *ith* gray wolf and X_p^{iter} represent the position vector of the prey (both position vectors are *d*-dimensional), c_a and c_p are coefficient vectors, and *iter* is the iteration number. The mathematical equation representing the gray wolf encircling operation is given by Equation 15.1.

$$X_i^{iter+1} = X_p^{iter} - c_a D_i^{iter} \tag{15.1}$$

where

$$D_i^{iter} = \left| c_p X_p^{iter} - X_i^{iter} \right|$$

The coefficient vectors are given by Equation 15.2:

$$c_a = 2a^{iter} r_1 - a^{iter} \text{ and} \tag{15.2}$$

where r_1 and r_2 are random vectors with values in the range [0, 1], and a^{iter} is a parameter (vector) whose component values decreases from 2 to 0 as the number of iterations increases. The parameter a^{iter} is defined as

$$a^{iter} = 2 - (iter) \frac{2}{MaxIter} \tag{15.3}$$

Figure 15.5 shows the diagrammatical model of a gray wolf encircling prey. In Figure 15.5 the dark circle represents the gray wolf at coordinates $[x_{i1}, x_{i2}]$ (dimension vector), and the dark triangle represents the prey at coordinates $[x_{p1}, x_{p2}]$ (dimension vector). The light circles (numbered as 1, 2, 3, 4, 5, 6, 7) represent possible positions around the prey that

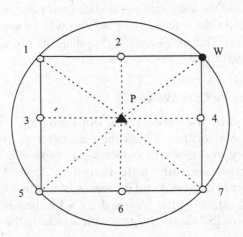

FIGURE 15.5
Modeling of gray wolf encircling prey.

could be taken up by other wolves of the pack. The various possible positions as given by Equation 15.1 could be attained by varying the values of c_a and c_p. The wolves can take up any position around the prey other than those indicated in Figure 15.5 by randomly varying the values for r_1 and r_2 in the interval [0, 1]. This two-dimensional representation of wolves encircling the prey can be extended to d-dimensions, thus forming a hypercube.

15.3.2 Hunting Behavior of Gray Wolves

The hunting of gray wolves is led by the *alpha* of the pack, sometimes along with *beta* and *delta* wolves. Assuming that these *alpha*, *beta*, and *delta* wolves have good knowledge of the location of the prey, they are taken as the best three solutions obtained so far in the search space. The optimum solutions are ordered according to the position of *alpha*, *beta*, and *delta* wolves respectively. The *omega* wolves take up or update their positions according to the position of *alpha*, *beta*, and *delta* wolves of their pack. The mathematical modeling of this behavior is given by the following equations:

$$X_{\omega 1}^{iter} = X_\alpha^{iter} - c_1 D_\alpha^{iter}, \, X_{\omega 2}^{iter} = X_\beta^{iter} - c_2 D_\beta^{iter}, \, X_{\omega 3}^{iter} = X_\delta^{iter} - c_3 D_\delta^{iter} \tag{15.4}$$

where $D_\alpha^{iter} = \left| c_\alpha X_\alpha^{iter} - X^{iter} \right|$, $D_\beta^{iter} = \left| c_\beta X_\beta^{iter} - X^{iter} \right|$, $D_\delta^{iter} = \left| c_\delta X_\delta^{iter} - X^{iter} \right|$

$$X_\omega^{iter+1} = \frac{X_{\omega 1}^{iter} + X_{\omega 2}^{iter} + X_{\omega 3}^{iter}}{3} \tag{15.5}$$

The *omega* wolves of the pack update their positions based on the positions of *alpha*, *beta*, and *delta* wolves as given in Equation 15.5. The diagram modeling the hunting behavior of wolves is given in Figure 15.6.

In this figure, P represents the prey and R is the radius of the circle at the center of which the prey is located. The *alpha*, *beta*, and *delta* wolves estimate and position themselves around the prey, and the *omega* wolves update their positions according to the positions of the *alpha*, *beta*, and *delta* wolves. The positions will be somewhere within a circle, which is randomly located. The *omega* wolf X_ω^{iter} updates its position to X_ω^{iter+1} while moving towards the prey, as given by Equation 15.5.

15.3.3 Attacking of Prey by Gray Wolves

When the prey, encircled by the gray wolves, stops moving, it is attacked. This is the exploitation phase in the algorithm. When the parameter c_a is decreased, the wolf is moving towards the prey, and when c_a is increased the wolf is moving away from the prey. The value of c_a varies randomly in the range $[-2a^{iter}, +2a^{iter}]$. The parameter a^{iter} is dependent on the number of iterations and decreases from 2 to 0 as the iterations increase. When $|c_a| < 1$, the wolf is attacking the prey, and $|c_a| > 1$ indicates that the wolf is moving away from the prey. Figure 15.7 shows the gray wolf attacking the prey and moving away from the prey.

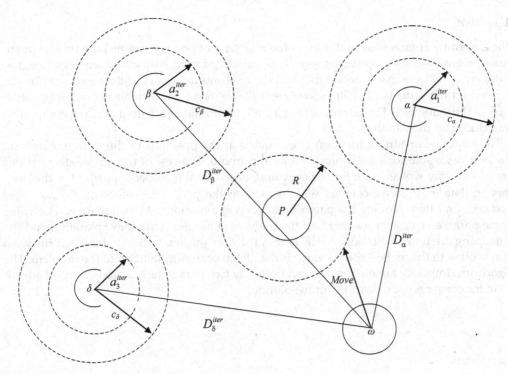

FIGURE 15.6
Mathematical modeling of hunting behavior of gray wolves.

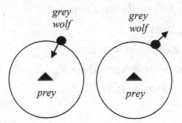

FIGURE 15.7
Modeling of gray wolves attacking prey.

15.3.4 Gray Wolves Searching for Prey (Exploration)

The gray wolves search for prey based on directions or indications from the *alpha, beta,* and *delta* wolves of the pack. This exploration is modeled by the parameter c_a where $|c_a| > 1$ indicates exploration of the search space. This is divergence of the gray wolves. Another parameter that models exploration is c_p whose value ranges in the interval $[0, 2]$. The values assumed by c_p are random, thus introducing a stochastic nature into the distance between prey and the wolf. This is also helpful in coming out of local optima. Since the position vector of the prey, denoted as X_p^{iter}, gets multiplied by the random component c_p, it acts as a weighting component for the position vector of the prey, making it harder for the wolf to approach the prey. This is also equivalent to having obstacles in the path between the wolf and the prey.

Algorithm

The algorithm commences with the problem statement and constraints, if any. The population of gray wolves (population size N) is initialized randomly in the search space. The d-dimensional objective function $f(X)$ and the maximum number of iterations *MaxIter* are defined for the problem. The iterations are indexed using the variable *iter* and it is initialized to the value of 1. The parameters c_a, c_p, a^{iter} are initialized with a random component introduced by the variables r_1 and r_2.

The objective or fitness function is evaluated at the positions of the gray wolves, and the best three solutions are designated as α, β, and δ. The rest of the candidate solutions are ω. The gray wolves hunt for the prey and, based on the probable position of the prey, they update their own positions with respect to the prey. The values of c_a, c_p, a^{iter} are updated, and they provide the balance between exploration and exploitation. The value of the parameter c_a ensures that half the number of iterations are for exploration and the remaining are for exploitation. At the end of the maximum number of iterations the *alpha* gray wolf with the highest fitness value is the global optimum solution to the problem. The algorithm simplicity arises from the fact that only two parameters, c_p and a^{iter}, need adjustment for convergence to the optimum solution.

Pseudocode

Initialization

Population of gray wolves X_i $i = 1, 2, ..., N$

Define objective function $f(X)$ with dimension d

Initialize parameters $a, c_a,$ and c_p

Calculate the fitness value of the population

Three highest fitness values are assigned to $X_\alpha, X_\beta, X_\delta$ respectively

Maximum number of iterations *MaxIter*

$iter = 1$

while ($iter \leq MaxIter$) **do**

  for $i = 1$ to N_ω

      Update position of X_i by Equation 15.5 (N_ω *is population size of omega wolves*)

  end for

  Update the parameters $a, c_a,$ and c_p

  Calculate the fitness values of the entire wolf population

  Update $X_\alpha, X_\beta, X_\delta$

  $iter = iter + 1$

end while

Highest fitness value X_α is the global optimum solution

Flowchart

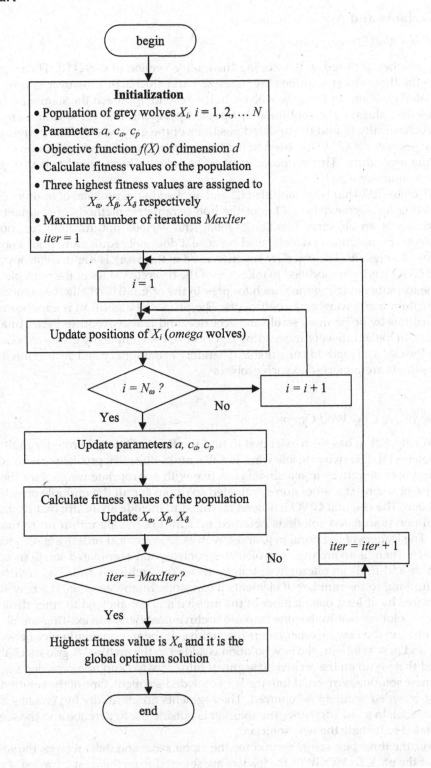

15.4 Variants and Applications

Binary Gray Wolf Optimization

Two approaches are used in developing the binary version of GWO [1]. The steps taken towards the three fittest solutions are binarized and stochastic crossover is used to find the updated positions of the gray wolves. In the second approach the sigmoidal function squashes the values of the continuously updated positions, and these values are thresholded stochastically to find the updated positions of the gray wolves. The performance of the binary version of GWO has been found to be superior for data classification compared to existing algorithms. This proposed method has always been found to converge to the global optimum solution.

The discrete GWO has been modified [2] and applied to the problem of multilevel image thresholding for segmentation. One of the non-parametric methods of segmentation is optimization of an objective function. Among the various optimization functions, the Kapur's entropy function is widely used because it does not require any prior knowledge and gives good results. The objective function used in this work is Kapur's entropy and the original GWO has been modified to take care of the discrete values of the multiple thresholds. The formula modeling the search for prey in the original GWO has been modified in this algorithm using weighting coefficients. The proposed algorithm is superior to existing algorithms by being more stable and accurate, and it also produces better image segmentation. In the multilevel thresholding algorithms, the complexity grows exponentially, making the classical methods unsuitable. Therefore evolutionary and swarm intelligence-based methods are preferred in such problems.

Multi-Objective Gray Wolf Optimization

The GWO algorithm has been extended to find the Pareto Optimal Front for multi-objective problems [3]. The two possible ways to solve multi-objective problems are to combine all the multiple objectives into a single objective with appropriate weights for the single objectives, or to find the set of non-dominated solutions for all the multiple objectives for the problem. The original GWO has been modified to include an *archive* that contains the set of all non-dominated solutions obtained by running the algorithm up to that point of time. The leaders of the hunt, α, β, and δ wolves (hierarchical order), are selected from the archive. There is an archive controller that controls non-dominated solutions entering the archive and finds an alternative strategy when the archive is full. The archive has a maximum limit to the number of elements it can store. In any iteration, if a new solution is dominated by at least one member of the archive it is not allowed to enter the archive. If the new solution dominates one or more archive members, the existing members are thrown out and the new member enters the archive. If there is no dominance between the existing and new solutions, the new solution is added to the archive. A grid mechanism is proposed that is run on the archive to segment into crowded and non-crowded segments, and the new solution is inserted into the least crowded segment. One of the solutions from the most crowded segment is removed. The segments are normally hypercubes that are *d*-dimensional. In a few instances, the solution is outside the hypercube and the segments are extended to include the new solution.

In GWO, the three best solutions become the alpha, beta, and delta wolves, the so-called leaders of the pack. In MOGWO, the leaders are selected from the least crowded segments of the solution hypercube based on a roulette wheel mechanism. The probability of selecting a leader from one of the segments (hypercube) is given by $p_k = \dfrac{c}{N_k}$ where p_k is the probability of selection, N_k is the number of members in the Pareto Optimal Front in the *kth* segment, and c is a positive constant with value greater than one. When the value of N_k is small, the segment is less crowded and the probability of selection is higher. The three leaders in the hierarchy (alpha, beta, and delta) are chosen from the least crowded hypercube if there are three or more members. If there are less than three members in the least crowded hypercube, the leaders are chosen from other hypercubes which are in the order of decreasing crowd.

The algorithm and pseudocode of the MOGWO are the same as those of the GWO algorithm except for the following differences:

- An archive is created to store non-dominated solutions and the leaders – alpha, beta, and delta are chosen from the archive whereas in GWO the leaders are chosen from the available population based on their fitness values.
- During each iteration, after updating the positions and objective function values of the wolf population, the archive is updated with the non-dominated solutions. If there is no space in the archive, some of the solutions are eliminated in order to accommodate the new non-dominated solutions. In case any of the new solutions are outside the archive, the hypercube is extended to include the new solutions.
- New leaders are selected from the updated archive using the same process as before.
- Finally, the result of the algorithm is the set of non-dominated solutions available in the archive whereas in GWO it is the alpha that is the optimum solution.

The parameters controlling the performance of MOGWO are the same as those of GWO. MOGWO has been found to be robust and stable and competitive with other multi-objective algorithms with good convergence. Feature selection for classification [4] is one of the important problems that has a far-reaching effect on the performance of the classifier. MOGWO applied for feature selection has been found to outperform GA and PSO. The training and operation phase of a classifier depends to a large extent on the feature set (attributes of the objects to be classified). Choosing the appropriate features leads to data reduction and elimination of redundancy to a certain extent. This speeds up the classifier and reduces the search space.

Another variant of the original GWO algorithm is [5] wherein approximately half the iterations are for exploration of the search space and the remaining half are for exploitation. This exploration/exploitation is balanced in this version of GWO in order to find the global optimum with the right balance of exploration/exploitation. This is achieved by modifying the parameter a^{iter} as follows:

$$a^{iter} = 2 - 2\frac{iter^2}{(MaxIter)^2} \tag{15.6}$$

This makes exploration of the search space 70% and exploitation 30% compared to 50–50 for the original algorithm. The algorithm has proved its superiority by solving the clustering problem in wireless sensor networks. The GWO algorithm has been employed to solve the node localization problem in wireless sensor networks successfully. GWO is found to converge faster with maximum accuracy or least localization error. The results are attributed to the explorative and exploitative abilities of the algorithm that avoids local optima and finds the global optimum in many cases [6].

The GWO algorithm is improved by introducing reinforcement learning along with neural networks [7]. Reinforcement learning is mainly used in selecting the right parameters for the GWO algorithm. The single parameter that decides whether the algorithm is exploring the search space or is in exploitation is the same throughout the algorithm in the original GWO. In this modified algorithm, the parameter is changed for each wolf separately using the reinforcement learning technique. The search space of each wolf and its own experience are used in fixing the parameter value that decides between exploration and exploitation. The repository of the experience of the wolves is built using neural networks that are updated by all the wolves in every iteration. The algorithm has established its superiority over GWO, PSO, and GA on the feature selection problem and designing optimal weights for neural networks.

An improvement in the GWO algorithm is proposed [8] with invasion-based migration operation. In normal GWO there is one pack of wolves that hunts for prey (search operation), encircles prey, and attacks prey. In this modified GWO there is more than one pack of wolves with migration of wolves between the different packs. This paves the way for new solutions based on information exchange. This invasion-based migration helps the algorithm in coming out of local optima. The algorithm is initialized with more than one pack of wolves in the search space. When the algorithm is stuck in local optima, the migration operation is deployed. The best pack of wolves (in terms of fitness) is chosen for the wolves from other packs to migrate to. The wolves that will migrate are chosen to be the best ones in their own pack. The best pack will have a higher number of wolves, and the ones with the lowest fitness values are eliminated. The other packs with lower numbers of wolves generate new wolves in a random manner. The performance of the modified GWO is found to be efficient in solving complex problems.

The chaos theory integrated into GWO [9] increases its speed of convergence. The algorithm is compared with standard metaheuristic algorithms including the original GWO and its performance is found to be better with the right chaotic map. Chaotic maps are used in finding the right value for the parameter a. The algorithm shows competitive performance with respect to other algorithms for constrained engineering optimization design problems, especially in convergence to the global optimum.

GWO has been applied [10] to tuning the parameters of Takagi–Sugeno proportional-integral fuzzy controllers (T–S PI-FCs). Here, the GWO algorithm has been applied to find the global minimum of the objective function formulated as the sum of the control error (absolute value) and the square of the output sensitivity function. The individual components of the function are weighted. The variables involved are the tuning parameters of the controller. The application is a non-linear servo system. GWO gives better performance than PSO and GSA.

An integrated GWO with kernel extreme learning machine (IGWO–KELM) is applied for finding the right feature set in medical data classification [11]. The performance of the algorithm with respect to accuracy, sensitivity, and specificity, which are used in evaluating classification problems, is found to be better than GA and GWO. This classification is important for the medical diagnosis of diseases based on data available.

The drawback of GWO is its inability to come out of local optima. This is overcome [12] by integrating GWO with the cuckoo search algorithm to find the global optimum. The GWO algorithm is modified by introducing cuckoo search in updating the alpha, beta, and delta wolves of the pack with each iteration. This improves the global exploration of the search space of the GWO algorithm. The GWO algorithm has been employed to solve the node localization problem in wireless sensor networks successfully.

15.5 Summary

The GWO algorithm has been modeled on the social hierarchy and behavior of gray wolves as a pack. Their leadership hierarchy, the subservience of the wolves to the leaders, and their activities undertaken for the benefit of the entire pack have been the inspiration behind the GWO algorithm. The hunting, chasing, encircling, and attacking operations on prey have been mathematically modeled in the algorithm and implemented successfully. The group behavior of wolves in obeying orders, foraging, and taking care of their pack is unique to the gray wolf family.

The performance of the GWO algorithm is competitive with particle swarm optimization, differential evolution, and the gradient search algorithm and even outperforms some of them for a few benchmark functions. The results are attributed to the explorative and exploitative abilities of the algorithm which avoids local optima and finds the global optimum in many cases. The abrupt changes in the trajectory of a search agent makes it explore the search space whereas reduction in the step size of the search agent enhances the exploitative abilities of the algorithm. An optimization algorithm should be balanced between these two characteristics in order to exhibit good convergence properties. The GWO algorithm is found to have good exploration, exploitation, avoidance of local optima, and convergence to the global optimum solution, and it is suitable for real-time applications.

References

1. E. Emary, Hossam M. Zawbaa, Aboul Ella Hassanien, Binary grey wolf optimization approaches for feature selection, *Neurocomputing*, Vol. 172, pp. 371–381, January 2016.
2. Linguo Li, Lijuan Sun, Jin Qi, Bin Xu, Shujing Li, Modified discrete grey wolf optimizer algorithm for multilevel image thresholding, *Computational Intelligence and Neuroscience* (Hindawi), Vol. 2017, pp. 1–17, 2017.
3. Seyedali Mirjalili, Shahrzad Saremi, Seyed Mohammad Mirjalili, Leandrodos S Coelho, Multi-objective grey wolf optimizer: A novel algorithm for multi-criterion optimization, *Expert Systems with Applications* (Elsevier) Vol. 47, pp. 106–119, April 2016.
4. E. Emary, Waleed Yamany, Aboul Ella Hassanien, Vaclav Snasel, Multi-objective gray-wolf optimization for attribute reduction, *International Conference on Communication, Management and Information Technology (ICCMIT 2015)*, *Procedia Computer Science*, Vol. 65, pp. 623–632, 2015.
5. Nitin Mittal, Urvinder Singh, Balwinder Singh Sohi, Modified grey wolf optimizer for global engineering optimization, *Applied Computational Intelligence and Soft Computing* (Hindawi), Vol. 2016, pp. 1–17, 2016.
6. R. Rajakumar, J. Amudhavel, P. Dhavachelvan, T. Vengattaraman, GWO-LPWSN: Grey wolf optimization algorithm for node localization problem in wireless sensor networks, *Journal of Computer Networks and Communications* (Hindawi), Vol. 2017, pp. 1–11, 2017.

7. E. Emary, Hossam M. Zawbaa, Crina Grosan, Experienced gray wolf optimization through reinforcement learning and neural networks, *IEEE Transactions on Neural Networks and Learning Systems*, Vol. 29, No. 3, pp. 681–694, March 2018.

8. Duangjai Jitkongchuen, Pongsak Phaidang, Piyalak Pongtawevirat, Grey wolf optimization algorithm with invasion-based migration operation, *2016 IEEE /ACIS 15th International Conference on Computer and Information Science (ICIS)*, Japan.

9. Mehak Kohli, Sankalp Arora, Chaotic grey wolf optimization algorithm for constrained optimization problems, *Journal of Computational Design and Engineering*, Vol. 5, No. 4, pp. 458–472, October 2018.

10. Radu-Emil Precup, Radu-Codrut David, Alexandra-Iulia Szedlak-Stinan, Emil M. Petriu, Florin Dragan, An easily understandable grey wolf optimizer and its application to fuzzy controller tuning, *Algorithms*, Vol. 10, No. 2:68, pp. 1–15, 2017.

11. Qiang Li, Huiling Chen, Hui Huang, Xuehua Zhao, ZhenNao Cai, Changfei Tong, Wenbin Liu, Xin Tian, An enhanced grey wolf optimization based feature selection wrapped kernel extreme learning machine for medical diagnosis, *Computational and Mathematical Models in Medicine* (Hindawi), Vol. 2017, pp. 1–15, 2017.

12. Hui Xu, Xiang Liu, Jun Su, An improved grey wolf optimizer algorithm integrated with cuckoo search, *9th IEEE International Conference on Intelligent Data Acquisition and Advanced Computing Systems: Technology and Applications (IDAACS)*, Bucharest, Romania, September 2017.

16

Elephant Herding Optimization

16.1 Introduction

The elephant herding optimization (EHO) algorithm is a swarm intelligence algorithm based on the herding behavior of elephants. The swarm intelligence-based algorithms are metaheuristic algorithms and provide solutions to complex real-world problems with reduced time complexity. These algorithms are modeled on the collective intelligence and social interactions among swarms of animals, birds, and insects. The solutions found by these metaheuristic algorithms might not be accurate, but they are close enough to the exact solution to be optimum. This is the main advantage of evolutionary as well as swarm-based algorithms. Since the particle swarm optimization (PSO) algorithm was first proposed in 1995, several swarm-based optimization algorithms have been invented, and the EHO is one such algorithm based on the herding behavior of elephants that was developed in 2015 [1, 2]. The myriad of problems that occur in real-life applications has motivated the development of nature-inspired optimization algorithms since the traditional methods do not produce satisfactory results for all kinds of problems. Evolution in nature has motivated the invention of several optimization algorithms, and these algorithms are either based on evolutionary principles and/or swarm intelligence.

Nature has its own methods of tackling problems, and this is obvious in the survival techniques undertaken by animals, birds, and insects. One such behavior is exhibited by elephants which are one of the largest mammals on earth. The young and old female elephants live together in clans, and these clans are headed by a female elephant called the matriarch. The study of the elephant clans with their disciplined behavior within the clan and the separation of male elephants from the clan has motivated the development of the EHO algorithm. Elephants are social animals, and their behavior has been modeled as two operators – *clan updating* and *clan separation*. This involves metaheuristics since nature is not deterministic and does not provide exact answers to problems. But most species have survived for generations because of their adaptability to nature and the dynamically changing hostile environment. The elephants in the clan update themselves with respect to their current position and the position of the matriarch. This characteristic of elephants is modeled as the clan *updating* operator. The male elephants form part of the clan till they reach puberty and then separate from the clan and live in isolation. This behavior is modeled by the clan *separating* operator. These two operations are mathematically modeled into the elephant herding optimization algorithm, and it has provided satisfactory results to the benchmark functions and standard test problems. EHO has shown excellent performance in solving intractable problems even though it has poor exploitation properties and slow convergence.

16.2 Elephant Herding Behavior

Elephants are one of the largest mammals found in Asia and Africa. The Asian and African elephants have several sub-species (two to four) amongst their population. Elephants are intelligent and have good memory and exhibit emotions like grief and joy, and can play around. Their long memory helps them in remembering the location and presence of watering holes that are quite a distance away. They have a long trunk that is used for drinking and eating. Elephants also lift things with their trunk. Elephants communicate over long distance using a subsonic vibration that travels over ground through air faster than sound. Male elephants maintain contact with their clan through low-frequency vibrations. These vibrations are detected through the skin on the feet and trunks. Figure 16.1 shows a picture of an elephant and its baby at the Chester Zoo, Cheshire, England.

Elephants form and live in related family groups. The group or clan size might go up to 50 members. The elephant population is organized into clans, and a clan consists of female elephants and their calves. Every clan is led by the oldest female elephant of the group called the matriarch. The male elephant usually separates from the clan once it is grown up and attains puberty, and lives in isolation. Within the group, the elephants express sympathy, loyalty, and cooperation. Elephants feed on a diet of sugarcane, bananas, bamboo, crops, grasses, coconut, etc. Climate changes have made their habitat hotter and drier leading to dwindling of their population. Poaching and superimposing on elephant habitats by humans have led to dwindling of elephant population and reduced forest area for elephants to occupy. Consequently, this leads to large-scale conflict with humans. Figure

FIGURE 16.1
Elephants in Chester Zoo. (Author: Nigel Swales. **Source: Flickr, CC BY-SA 2.0. https://creativecommons.org/licenses/by-sa/2.0/deed.en.**)

FIGURE 16.2
Elephants crossing the Luangwa River in Zambia. (Author: Geoff Gallice. **Source: Flickr, CC BY 2.0. https://cr eativecommons.org/licenses/by/2.0/deed.en.)**

16.2 shows a herd of elephants crossing the Luangwa River in South Luangwa National Park, Zambia.

16.3 Elephant Herding Optimization

The EHO algorithm is based on the behavior of elephant herds. Some simplifying assumptions have been made in the development of the EHO algorithm. It is assumed that elephants are composed of clans with a fixed number of elephants in each clan. All the elephants of a clan live under the leadership of a matriarch which is the oldest female elephant of the entire group. The elephants update their own positions based on the position of the matriarch in every iteration. In addition, one male elephant will leave the clan to live separately. The characteristics of elephants in living together as clans under the leadership of a matriarch and the separation of male elephants from the clan when they reach puberty have been modeled into the optimization algorithm. This behavior of elephants is modeled as two operators in the algorithm – clan *updating* and *separating*.

16.3.1 Algorithm

To simplify the development of the algorithm, it is assumed that the number of clans is N_C and the number of elephants in every clan is the same and equal to N_{CE}. It is also assumed that only the worst elephant will leave the clan with every iteration. The variable i is used as an index for the clans, and the variable j is used for indexing the elephants within the clan. Therefore, $i = 1, 2, ..., N_C$ and $j = 1, 2, ..., N_{CE}$. The search space is assumed to be of dimension d and is indexed by the variable k, that is, $k = 1, 2, ..., d$. The maximum number of iterations in the algorithm is given by the variable *MaxIter*, and the iterations are indexed by the variable *iter*. The position of every elephant in the search space in iteration *iter* is represented by

$$X_e^{iter}(i,j) = \left[x_{e1}^{iter} \ x_{e2}^{iter} \ ... \ x_{ed}^{iter} \right] \tag{16.1}$$

The positions of elephants in each clan are *updated* based on their current position and the position of the matriarch of the clan. The influence of the matriarch of the clan is also included in the updating process of the position of the elephants. The worst elephant is replaced using the *separating* operator. The *separating* operation leads to population diversity.

The position of every elephant *j* in the clan *i* is *updated* by Equation 16.2 (except the matriarch):

$$X_e^{iter+1}(i,j) = X_e^{iter}(i,j) + s_m.rand.\left[X_e^{iter}(i,best_i) - X_e^{iter}(i,j) \right] \tag{16.2}$$

where $X_e^{iter}(i,j)$ is the current position (in iteration *iter*) of one elephant in clan *i*, index *j* is the *j*th elephant within clan *i*, $best_i$ is the best elephant of the clan *i* which is the matriarch of the clan, s_m is a scale factor in the range [0, 1] that represents the influence of the matriarch on the elephant *j* in the clan, and *rand* is a random number uniformly distributed in the interval [0, 1]. Similarly $X_e^{iter+1}(i,j)$ is the updated position of the *j*th elephant in the *i*th clan.

The position of the fittest elephant (matriarch) of the clan is updated by the following equation:

$$X_e^{iter+1}(i,best_i) = r.X_c^{iter}(i) \tag{16.3}$$

where *r* takes values in the interval [0, 1] and represents the influence of the center of the clan *i* on the fittest matriarch elephant of the clan, and X_c^{iter} is the center position of the clan in iteration *iter*. The dimension of the search space is *d*, and let the index for the dimension be *k*, that is, *k* = 1, 2, ..., *d*. The center of the clan is computed as follows for the *k*th dimension in the *d*-dimensional space (Equation 16.4):

$$X_c^{iter}(i,k) = \frac{1}{N_{CEi}} \sum_{j=1}^{N_{CEi}} X_e^{iter}(i,j,k) \tag{16.4}$$

Equation 16.4 is repeated *d* times for the *d*-dimensions of the search space (*k* = 1, 2, ..., *d*).

The male elephants leave their clan and live alone solitarily when they reach puberty. This behavior of male elephants is modeled using a *separation* operator in the EHO algorithm. The equation to implement this *separation* operation is given by Equation 16.5:

$$X_e^{iter}(w_i,i) = X_e^{min} + (X_e^{max} - X_e^{min} + 1).rand \tag{16.5}$$

where w_i is the worst elephant in the clan i, $X_e^{\min}$ is the lower bound on the elephant position and $X_e^{\max}$ is the upper bound on the elephant position, and *rand* is a random number with a uniform distribution that takes values in the range [0, 1].

The parameters in the clan *updating* and *separating* operations that take random values in the interval [0, 1] have a uniform distribution in the specified range. These random values and the clan separating operator improve the diversity of the population.

16.3.2 Pseudocode

Initialization

    Initialize number of elephant clans N_C

    Initialize number of elephants in each clan N_{CE}

    Initialize the number of dimensions d in the search space

    Define the objective function $f(X)$

    Compute the fitness values of all the elephants in all the clans

    Initialize the maximum number of iterations *MaxIter*

    iter = 1

while (*iter* ≤ *MaxIter*)

    Sort the elephant population according to fitness values

    clan updating operator

      for i = 1 to N_C

        for j = 1 to $N_{CE}(i)$

          if ($j \neq best$)

            Update position of elephant j in clan i (Equation 16.2)

          elseif ($j = best$)

            Update position of fittest elephant (matriarch) in clan i (Equation 16.3)

          end if

        end for

      end for

    clan separating operator

      for i = 1 to N_C

        Apply the separating operator on the worst elephant of clan i (Equation 16.5)

      end for

    Evaluate the elephant population in their new updated positions

    iter = *iter* + 1

end while

Elephant with the best fitness among the entire population is the global optimum

Flowchart

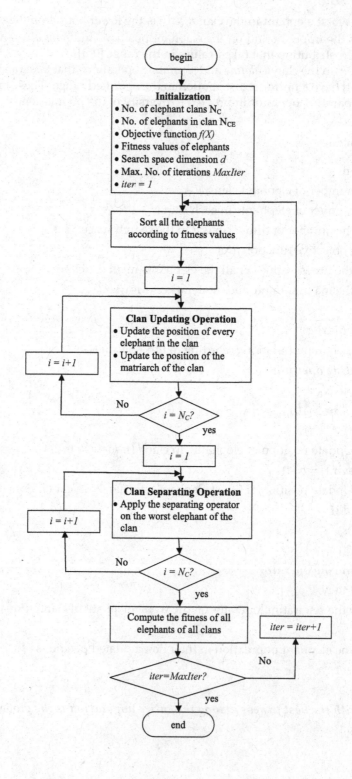

16.4 Variants of the Algorithm

The **enhanced elephant herding optimization** (EEHO) algorithm was proposed as a variant of EHO to provide a good balance of exploration and exploitation and prevent premature convergence towards the origin. The algorithm has a new position updating operator that controls the convergence towards the center of the clan and the matriarch and enhances the performance and also maintains diversity of population [3].

The elephants update their positions with each iteration depending on their current position and the influence of the matriarch on the elephant being updated. But one of the disadvantages is that the updated position need not always be better than the current position (fitness value). In [4] three different enhancements to EHO have been proposed – culture-based, alpha tuning, and biased initialization EHO. Alpha tuning improves the performance by continuously varying the scale factor with every iteration. The scale factor s_m changes linearly with an increasing number of iterations instead of having a constant value for s_m. s_m represents the influence of the matriarch on the elephants of the clan. In culture-based EHO, the worst elephant is replaced with a saved elephant from the belief space that has higher fitness. The belief space consists of elephants with higher fitness values. Another proposed strategy is to initialize the elephant population with good fitness values. A threshold can be defined for this fitness. This is the biased initialization and the evolution starts only after the fitness reaches a threshold level.

In [5] the EHO has been tested on 28 CEC2013 benchmark functions and found to be comparable in performance to PSO.

Elitism: It is a strategy adopted by the EHO algorithm where the worst elephants are replaced by the saved elephants that have good fitness values. The saving of elephants takes place after evaluation of the fitness and sorting of the elephants based on these fitness values and before the application of the clan updating operator. The clan updating operator changes the positions of the elephants and hence their fitness values since the fitness depends on the position. By this variation in the EHO algorithm the elephants with high fitness values are preserved; normally one or two elephants are saved. Moreover, the elitism strategy adopted also improves the rate of convergence of the algorithm.

16.5 Summary

Nature has taught us that the collective intelligence and social interactions among the members of a group is able to accomplish much more than individuals can accomplish. The intelligence and smart behavior exhibited by elephants have been the source of inspiration for development of the EHO algorithm. It has been successful in providing solutions to complex engineering problems that are quite challenging and time-consuming to solve by other traditional methods. Typical values for the parameters of the EHO algorithm are $N_C = 5$, $N_{CE} = 10$, and this makes the total elephant population size 50. The scale factor for influence of the matriarch on the clan updating operation is $s_m = 0.5$, and the factor $r = 0.1$ is the influential parameter for matriarch updating. The parameter values and the in-built stochasticity in the algorithm lead to differences in results with each run of the algorithm, but on average, the optimum result is obtained. The dimension of the population space could be between 10 and 20 and the maximum number of iterations could range from 50 to 100, depending on the problem to be solved. The EHO algorithm has fewer parameters to control compared to other swarm-based algorithms. If the parameters are varied adaptively, especially as the iterations progress, the performance could improve as compared to having fixed parameter values for all the iterations of the algorithm. The EHO algorithm is easy to implement since it has a lower number of parameters compared to other optimization algorithms. The disadvantages are its slow convergence and lesser exploitation abilities. EHO has been tested on various benchmark problems and found to be competitive with the existing optimization algorithms such as the genetic algorithm (GA) and differential evolution (DE). EHO can be hybridized with other swarm intelligence algorithms to further improve the performance.

References

1. G.-G. Wang, S. Deb, X.-Z. Gao L. dos Santos Coelho, Elephant herding optimization, *IEEE 3rd International Symposium on Computational and Business Intelligence* (ISCBI), Bali, Indonesia, 2015.
2. G.-G. Wang, S. Deb, X.-Z. Gao, L. dos Santos Coelho, A new metaheuristic optimization algorithm motivated by elephant herding behaviour, *International Journal of Bio-Inspired Computation*, Vol. 8, No. 6, pp. 394–409, 2016.
3. Alaa A. K. Ismaeel, Islam A. Elshaarawy, Essam H. Houssein, Fatma Helmy Ismail, Aboul Ella Hassanien, Enhanced elephant herding optimization for global optimization, *IEEE Access*, Vol. 7, pp. 34738–34752, March 2019.
4. Mostafa A. Elhosseini, Ragab A. El Sahiemy, Yasser I. Rashwan, X. Z. Gao, On the performance improvement of elephant herding optimization algorithm, *Knowledge Based Systems* (Elsevier), Vol. 166, pp. 58–70, February 2019.
5. Viktor Tuba, Marko Beko, Milan Tuba, Performance of elephant herding optimization algorithm on CEC2013 real parameter single objective optimization, *WSEAS Transactions on Systems*, Vol. 16, pp. 100–105, 2017.

17

Crow Search Algorithm

17.1 Introduction

Engineering design problems for real-life applications can be quite complex in nature with many different feasible solutions. The number of parameters or decision variables involved could be quite large, and all possible values of these decision variables lead to numerous solutions. In optimization, the goal is to find the best among all of them which is the global optimum for the design problem. The multiple feasible solutions could be equivalent to the local optima of the objective function defined for the problem. Conventional optimization methods do not perform well in such cases, and many of them require computation of the derivatives of the objective function that might be a continuous function of variables. In choosing the best among the feasible solutions in problems involving several constraints, the time complexity becomes an important factor. When the number of design variables is large or the function is complex and there are many local optima, conventional methods fail to find the global optimum in finite time and are not efficient. The traditional methods work well when the design variables are smaller in number and the objective function is unimodal. For problems which do not meet these requirements, metaheuristics plays a vital role in finding the global optimum solution in finite time.

Researchers have turned to nature for inspiration, and many optimization algorithms have been developed based on the study of the behavior of animals, birds, and insects. Such nature-inspired algorithms incorporate some randomness within them, and almost all of them are metaheuristic. Metaheuristic algorithms have been found to be suitable for solving engineering optimization problems that have non-linear objective functions and constraints and are also multimodal. Metaheuristic algorithms balance between randomness (diversification) and local search (intensification); hence they are effective in solving multimodal, non-linear, complex optimization problems. One such metaheuristic algorithm based on the intelligent behavior of crows is the *crow search algorithm* (CSA) discussed in this chapter [1]. Crows are very intelligent birds that exhibit flocking behavior, and they have been the source of inspiration behind the crow search optimization algorithm.

17.2 Crows in Nature

Crows are medium-sized black birds found commonly in several countries. Crows are aggressive birds, and they live in grasslands and fields with trees in the neighborhood where they can build nests. There are several species of crows (~40) around the world, and

FIGURE 17.1
Female house crow in Kuala Lumpur. (Author: Gerifalte Del Sabana – own work, CC BY-SA 4.0. https://creativ
ecommons.org/licenses/by-sa/4.0/deed.en.)

one common species called the *house crow*, typically found in most parts of Asia, is shown
in Figure 17.1. Some of them have a grey neck while others are completely black.

Crows, ravens, and rooks are all part of the genus *Corvus* that belongs to the *Corvidae*
family.

Ravens are bigger than crows with heavier bills and hoarser voices (sounds like a croak).
An Australian raven is shown in Figure 17.2. Rooks are slightly smaller than crows with
wedge-shaped tails and light-colored beaks, and one member of the species is shown in
Figure 17.3. They also have a harsh caw similar to that of the common crow.

Crows never forget a face. They can fiercely attack humans and other birds when their
territory is encroached upon or they feel threatened or to protect their young ones. They are
adaptable birds with a loud 'caw' that can be annoying and sound harsh to the ears. They
also have the honor of representing our dead ancestors, and feeding a crow is a custom
to pay respects to our earlier generations and receive blessings. Crows are very intelligent
birds, and they do steal food. They have excellent memory and search the environment for

FIGURE 17.2
Australian raven (*Corvus coronoides*), Doughboy Head, New South Wales, Australia. (Author: J. J. Harrison –
own work, CC BY-SA 4.0. https://creativecommons.org/licenses/by-sa/4.0/deed.en.)

FIGURE 17.3
Rook at Slimbridge Wetland Centre, Gloucestershire, England. [Author: Adrian Pingstone – own work (Public Domain).]

quality food sources in a manner equivalent to searching for the optimum solution in the search space. They forage either singly or in groups. When they locate a food source, they also communicate with other crows about their discovery of food by cawing to attract the attention of other crows. A group of crows is called a *murder*. When there is a death of a crow, they all join together and issue loud harsh cawing sounds. They mourn their dead and even investigate the cause of death. They also mob and chase away predators. Figure 17.4 shows crows in flight mobbing a red-tailed hawk.

Some species of crows live in roosting communities. A few species migrate to warmer climates, and many of them eat agricultural pests, whereas some of them damage crops and eat them. Sometimes they feed on dead animals and birds and garbage food. During

FIGURE 17.4
Crows mobbing a red-tailed hawk. (Author: Dori – own work, GFDL Ver 1.2. https://commons.wikimedia.org/wiki/Commons:GNU_Free_Documentation_License,_version_1.2.)

FIGURE 17.5
Crow's nest in Moscow. (Author: Bugaga – own work, GFDL Ver 1.2. https://commons.wikimedia.org/wiki/
Commons:GNU_Free_Documentation_License,_version_1.2.)

the mating period they build nests more than 15 feet above the ground. Female crows lay
4 to 5 eggs, and the chicks are hatched after an incubation period of 18 days. The nest of
a crow built in a tree containing four eggs is shown in Figure 17.5. The chicks are fed for
nearly 60 days. The lifespan of crows is approximately 14 years.

Crows are intelligent birds that have good memories for faces, communicate with other
members belonging to their species, and steal and hide food. It is remarkable that they
remember the hiding places of their food and protect their food from other birds and animals. They also change the hiding place, if necessary. When a crow finds that it is being
followed by another crow, it tries to fool the one following it in order to protect its food
stored in the hiding place. Crows have been found to have a large brain size compared to
their body size. They are very intelligent and can remember the hiding place of their food
even after several months. This behavior of crows is modeled in the *crow search algorithm*.

17.3 Crow Search Optimization

In the crow search algorithm, the environment is the search space, crows are the agents
searching for the solution (foraging for food), each position in the environment is a feasible
solution, the quality of the food in a location is related to the objective function value, and
the best food source found by the crow is the global best solution to the optimization problem. The four underlying principles of the crow search algorithm are:

- Crows exhibit flocking behavior (they live together).
- Crows remember the hiding place of their food.
- Crows follow other crows to steal food.

- Crows protect their own hiding places from being pilfered.

Assume that the environment is d-dimensional, and let the population size of crows be N (agents searching for solution in the search space). Let $X_i = [x_{i1} \ x_{i2} \ ... \ x_{id}]$, $i = 1, 2, ..., N$, be the vector representing the position of the ith crow in the d-dimensional search space where the dimension is indexed by the variable $j = 1, 2, ..., d$. The iterations are indexed by the variable $iter$, and the maximum number of iterations in the algorithm is $MaxIter$, where $iter = 1, 2, ..., MaxIter$. During any iteration represented by $iter$, the position of the ith crow in the d-dimensional space is given by the vector $X_i^{iter} = \left[x_{i1}^{iter}, \ x_{i2}^{iter}, ... x_{id}^{iter} \right]$. The hiding place of the ith crow during iteration $iter$ is represented by H_i^{iter}. Therefore this could be the best position attained by the ith crow during the search in the environment up to the current iteration $iter$. Crows keep looking for better food sources and hiding places in the environment.

Let H_k^{iter} be the hiding place of crow k during iteration $iter$ that it decides to visit. Crow i is following crow k and there are two possibilities: (i) crow k is not aware that crow i is following it, (ii) crow k is aware that crow i is following it.

(i) Crow i approaches the hiding place of crow k (crow k is unaware that crow i is following it) and the new position of crow i is given by:

$$X_i^{iter+1} = X_i^{iter} + r_i \ l_{fi}^{iter} \times \left(H_k^{iter} - X_i^{iter} \right) \qquad (17.1)$$

where r_i is a random number uniformly distributed in the range [0, 1], and l_{fi}^{iter} is the flight length of crow i during iteration $iter$ which could be lesser than or greater than 1. The above conditions of flight length less than or greater than 1 are diagrammatically represented in Figure 17.6 and Figure 17.7 respectively.

When the flight length l_{fi}^{iter} is less than one, the search is local and in the vicinity of the current position of crow i. The next position of crow i will be on the dotted line between

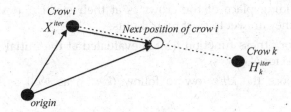

FIGURE 17.6
$l_{fi}^{iter} < 1$

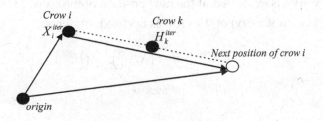

FIGURE 17.7
$l_{fi}^{iter} > 1$

X_i^{iter} and H_k^{iter}. When the flight length l_{fi}^{iter} is greater than one, the search becomes global and moves away from the current position of crow X_i^{iter}. The next position of crow i is again on the dotted line between X_i^{iter} and H_k^{iter} but it could exceed (go past) H_k^{iter}.

(ii) Crow k knows that crow i is following it and tries to fool crow i by going to a place different from where it has hidden the food. This is done so that crow i will not steal the food of crow k.

The two possibilities discussed above could be mathematically represented together as

$$X_i^{iter+1} = X_i^{iter} + r_i l_{fi}^{iter} \times \left(H_k^{iter} - X_i^{iter} \right) \text{ if } r_k \geq p_k^{iter}$$

$$= \text{random position,} \quad \text{otherwise}$$

(17.2)

where r_i and r_k are random numbers uniformly distributed in the interval [0, 1] and p_k^{iter} is the awareness probability of crow k in iteration *iter*. If the awareness probability has smaller values, the search becomes local and intensified, whereas if the awareness probability has larger values, the search becomes global and diversified.

17.3.1 Algorithm

The problem statement and associated constraints are defined. The decision variables and parameters such as population size N, flight length l_f, stopping criteria, maximum number of iterations *MaxIter*, and awareness probability p are identified.

1. The position and memory of the crows are randomly initialized in the search space. For example, the position of the *ith* crow (in iteration *iter*) is a d-dimensional vector defined as, $X_i^{iter} = [x_{i1}^{iter}, \ x_{i2}^{iter}, \dots \dots x_{id}^{iter}] \ i = 1, 2, \dots, N$.

   The memory (hiding place) of the crows is also a d-dimensional vector defined as $H_i^{iter} = [h_{i1}^{iter}, h_{i2}^{iter}, \dots \dots h_{id}^{iter}] \ i = 1, 2, \dots, N$

   Initially the hiding place of the crows is at their starting position, since it is assumed that they are yet to start their flights.

2. The objective or fitness function $f(X_i)$ is evaluated at the initial position of each crow (evaluated for N crows).

3. Randomly choose the *kth* crow to follow ($k \neq i$) as well as the parameters: $r_i, r_k, l_{fi}^{iter}, p_k^{iter}$.

4. The new position of the crow is generated according to Equation 17.1. The crows move to their new positions, if feasible; otherwise they remain in their present position.

5. The fitness values is evaluated at the new position of the crow.

6. The hiding place (memory) of the crow is updated as follows:

$$H_i^{iter+1} = X_i^{iter+1} \quad if \ f\left(X_i^{iter+1} \right) > f\left(H_i^{iter} \right)$$

$$= H_i^{iter} \quad otherwise$$

(17.3)

7. Steps 3 to 6 are repeated for all the N crows.

8. The algorithm terminates if either the stopping criterion is attained or the maximum number of iterations is reached; otherwise it continues.

9. The best (highest) fitness value evaluated among the hiding places of the N crows (as memorized by the crows) in the search space is the global optimum solution.

17.3.2 Pseudocode

Initialization

Population size of crows N

Objective (fitness) function $f(X)$ that is d-dimensional

Random initial position of the N crows which is also their initial hiding place (memory)

Evaluate fitness function at the initial positions of the crows

Define stopping criteria, if any

Maximum number of iterations *MaxIter*

$iter = 1$

while ($iter \leq MaxIter$) **do**

    for $i = 1 : N$

        Randomly choose the kth crow to follow

        Choose awareness probability p_k^{iter} for the kth crow in the current iteration

        Randomly choose a value for r_k in the interval [0, 1]

        if $r_k > p_k^{iter}$

$$X_i^{iter+1} = X_i^{iter} + r_i l_{fi}^{iter} \times \left(H_k^{iter} - X_i^{iter} \right)$$

    else

        X_i^{iter+1} is a random position in the search space

    end if

    end for

    Evaluate the fitness values at the new positions of the crows, if feasible

    Update the hiding places (memory) of the crows

    if stopping criterion is met, **exit**

    otherwise continue

    $iter = iter + 1$

end while

Highest fitness value among the memorized hiding places of the N crows is the global optimum

Flowchart

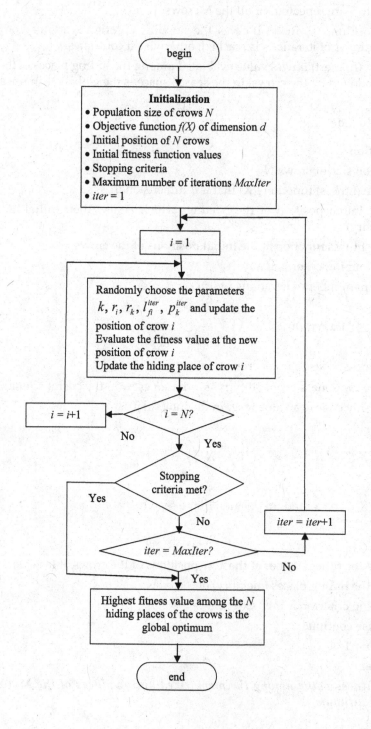

17.4 Variants and Applications

A hybrid algorithm based on the CSA and the sine cosine algorithm (SCA) called the sine cosine crow search algorithm (SCCSA) has been proposed [2]. The SCA uses sine cosine operators to update the positions of the particles in the search space. The disadvantages of the CSA, such as that the particles do not update themselves based on the global best and the particles move to new random positions based on flight length, are overcome in this hybrid algorithm. The hybrid algorithm has been found to be competitive with the other standard algorithms. CSA has been applied along with k-means clustering algorithm [3] to find the optimal number of clusters. The k-means algorithm is one of the most popular clustering algorithms in the literature, but its disadvantage is that it can get trapped in local minima. In this application, initially the k-means algorithm is applied to generate the cluster centers and then the *CSA-k-means* is applied to cluster the data around the centers so that optimal clustering is achieved. In [4] the CSA is applied to find the optimum conductor size in a radial distribution network under the constraints of the bus voltage limit and current carrying capacity of the conductor. The objective is to minimize the conductor capital cost and energy loss cost. This is helpful in choosing conductor type and size in the radial distribution network. It is found that the network power loss and annual cost are reduced compared to the original network. In [5] the CSA has been modified and applied to induction motors and distribution networks. The parameter identification in induction motors and capacitor allocation in distribution networks are two complex, non-linear, multimodal problems in the field of energy. Metaheuristic algorithms are promising for solving such problems in an optimal manner. Here CSA has been modified in terms of awareness probability and random perturbation to improve the diversity and convergence to the optimal solution. The awareness probability is made dynamic, i.e., it changes with respect to the fitness value of the candidates. Instead of a random number with a uniform distribution, Levy flights are used in this work which improves the efficiency in diversification. The problems chosen here are the internal parameters of an induction motor and the capacitor allocation in a distribution network that minimize the energy losses and improve the voltage profile.

Classical algorithms are time-consuming for solving fractional optimization problems (with a higher number of dimensions) since they rely on some transformation, which is overcome in metaheuristic algorithms. In [6] chaos theory is integrated into CSA to improve the convergence speed and improve diversification and intensification. Chaos theory tunes the parameters of the CSA to develop variants of the basic algorithm. The quality and reliability of the proposed algorithm in terms of global optimum solution, convergence speed, and computational time consumed surpass the other algorithms, especially for fractional optimization. The problem of job scheduling in industries or any other sector demands an optimum solution. The task becomes difficult when the number of jobs and/or resources is high. This problem has been tackled in [7] with the CSA, and the results are found to be better comparatively in terms of solution attained and time taken. In another research work [8] the CSA has been applied for local search in the two-stage algorithm of eagle strategy. In the first stage, global exploration with Levy flight takes place and if a promising solution is found, the search is intensified locally. For the local search another optimization algorithm is employed. In this paper the CSA is employed for local search to solve the unit commitment problem in smart grid systems. The optimization problem is to schedule the turning on/off of a set of generators to meet the load and operational constraints so that the total cost is minimized.

17.5 Summary

The crow search algorithm that is based on the intelligent behavior of crows has been explained in detail. It is a population-based algorithm that searches for the optimum solution in the search space. The different parameters associated with the algorithm that need to be tuned to obtain better solutions have been identified. The performance of the algorithm depends on the two important parameters – flight length and awareness probability. In addition, some randomness is introduced in the algorithm by the two random parameters r_i and r_k that follow uniform distribution. The awareness probability is an important parameter associated with CSA that provides the necessary balance between intensification and diversification. The CSA and its proposed variants applied for solving various engineering optimization problems that are available in the literature have also been outlined. This simple algorithm has been found to be effective in solving global optimization problems that are NP-hard in terms of convergence speed and computational complexity. CSA has been found to give good performance when compared to GA and PSO. CSA is robust, and optimum results can be obtained in less than 50 iterations on average. There are only two main parameters to be tuned for the algorithm which is an added advantage. Typical values for the population size are between 20 and 40, awareness probability is <0.2, and flight length is <2. CSA is simple and easy to implement and gives good performance over a wide spectrum of complex optimization problems.

References

1. Alireza Askarzadeh, A novel metaheuristic method for solving constrained engineering optimization problems: Crow search algorithm, *Computers and Structures* (Elsevier), Vol. 169, pp. 1–12, 2016.
2. Seyed Hamid Reza Pasandideh, Soheyl Khalilpurazari, Sine cosine crow search algorithm: A powerful hybrid meta heuristic for global optimization, *Third International Conference on Artificial Intelligence and Soft Computing*, August 2017.
3. Alireza Balavand, Ali Husseinzadeh Kashan, Abbas Saghaei, Automatic clustering based on crow search algorithm-Kmeans (CSA-Kmeans) and data envelopment analysis (DEA), *International Journal of Computational Intelligence Systems*, Vol. 11, No. 1, pp. 1322–1337, 2018.
4. Almoataz Y. Abdelaziz, Ahmed Fathy, A novel approach based on crow search algorithm for optimal selection of conductor size in radial distribution networks, *Engineering Science and Technology, an International Journal* (Elsevier), Vol. 20, pp. 391–402, 2017.
5. Primitivo Diaz, Marco Perez-Cisneros, Erik Cuevas, Omar Avelos, Jorge Galvez, Salvador Hinojosa, Daniel Zaldivar, An improved crow search algorithm applied to energy problems, *energies*, Vol. 11, No. 3, pp. 571, 2018.
6. Rizk M. Rizk-Allah, Aboul Ella Hassanien, Siddhartha Bhattacharyya, Chaotic crow search algorithm for fractional optimization problems, *Applied Soft Computing* (Elsevier), Vol. 71, pp. 1161–1175, 2018.
7. Antono Adhi, Budi Santosa, Nurhadi Siswanto, A meta-heuristic method for solving scheduling problem: Crow search algorithm, *International Conference on Industrial and System Engineering (IConISE)*, 2017.
8. Rachid Habachi, Achraf Touil, Abdelkabir Charkaoui, Abdelwahed Echchatbi, Eagle strategy based crow search algorithm for solving unit commitment problem in smart grid system, *Indonesian Journal of Electrical Engineering and Computer Science*, Vol. 12, No. 1, pp. 17–29, October 2018.

18

Raven Roosting Optimization Algorithm

18.1 Introduction

Nature-inspired optimization algorithms have been developed from the behavioral study of swarms of animals, birds, and insects and the biological evolution process, in general. Animals, birds, and insects interact with each other and also with the environment. This interaction is usually complex and leads to productive outcomes for the members of the swarm in the areas of foraging, mating, and protection against hostile elements. The animals, birds, and insects forage for food by applying a strategy that is unique to every species. The backbone of these foraging activities is the exchange of information among the members of the flock by gathering together. This becomes a social foraging activity, and each species has its own method of communication with other members of the flock. In most of the species, the members come home to roost after sunset and gather together. In the roost the information exchange takes place and members are recruited for foraging the next day at dawn. Roosting might take place either for one night or sometimes it might continue for days. Group foraging is very important for survival and protection against predators. The social roosting and foraging behavior of the common raven is the inspiration behind the development of the raven roosting optimization (RRO) algorithm discussed in this chapter [1]. Novel search strategies are required to find quality food by searching in the environment since the exact locations of food sources are not known in advance. Resources are available in plenty, but efficient and effective search strategies are required to find the quality food sources. This is essential for survival of the species, and most of them are adaptable in the dynamically changing, hostile environment. Darwin's theory of *survival of the fittest* has been proven by these species because those with good foraging strategies usually survive the evolution.

The roosting and foraging behavior of the common ravens has been a great source of inspiration for researchers of nature-inspired optimization algorithms. Several studies have been undertaken on the roosting and foraging behavior of the ravens and are available in the literature. The sharing of information about finding food is one of the important activities of the group. The method by which this information is communicated to other members of the group forms an important component of the optimization algorithm. It is found that ravens prey on dead animals and carcasses and the number of ravens at the foraging site varies from day to day until the prey is consumed completely. They leave the roost at dawn, prey on the food source, and return to the roost at sunset. This pattern of the ravens is repeated every day until the current food source is depleted. Then the group of ravens looks out for other food sources and once one is found, this is advertised to the other ravens in the roost. They follow the raven that has discovered the food source and share the food. This process is repeated except that the leader raven (the one who discovers the food) is different for each food source and the number of ravens that follow the leader also varies with each day. The optimization algorithm that is designed based on this raven roosting behavior has been described in detail in the following sections. These powerful swarm intelligence optimization algorithms have been applied to a wide range of applications and have proved to be efficient. They have been applied to a set of the standard benchmark functions and engineering design problems and have been found to produce results that are competitive with or better than the traditional algorithms.

18.2 Raven Roosting Behavior

The common raven (*Corvus corax*) belonging to the family *Corvidae* is an all-black bird and is one of the most intelligent birds found in nature. It is found widely distributed in the Northern Hemisphere, and there are different subspecies of the raven. The ravens have a habit of communal roosting [2], and their roosts are usually found in trees, bushes, abandoned buildings, coastal cliffs, and lakesides [3]. The exact location of the roost varies within the area depending on the presence of humans, animals, weather, and other similar factors. The roosts are maintained by ravens for years, and hundreds of ravens roost together. Ravens fly to the roosts nearing sunset, typically in a string formation similar to crows. Hundreds to thousands of ravens usually roost together, especially during the non-breeding months of the year. Figure 18.1 shows a picture of a common raven in Cypress Provincial Park, British Columbia, and Figure 18.2 shows a little raven in Australian National Botanic Gardens, Canberra, Australia. Figure 18.3 shows a North American common raven in majestic flight at Muir Beach in Northern California.

Roosts are nests where the birds come to rest usually at the end of the day, typically at sunset. Communal roosts function as centers [4] where information is exchanged and knowledge is gained. Ravens leaving the roosts normally move in a certain direction

FIGURE 18.1
Common raven in Cypress Provincial Park, British Columbia. (Author: User Clayoquot – own work, CC BY-SA 3.0. https://commons.wikimedia.org/wiki/Commons:GNU_Free_Documentation_License,_version_1.2; https://creativecommons.org/licenses/by-sa/3.0/deed.en.)

FIGURE 18.2
Australian raven (*Corvus coronoides*). (Author: J. J. Harrison – own work, CC BY-SA 3.0. https://creativecomm ons.org/licenses/by-sa/3.0/deed.en.)

together with several members. The direction in which the ravens leave the roost could vary with each day. The ravens that have acquired knowledge of the food source arrive at the location of food, accompanied by other birds from the roost. The raven that leads becomes the leader and the other ravens are followers, but the roles change based on the knowledge acquired about a new food source. Any bird in the roost is able to go to a

FIGURE 18.3
North American common raven (*Corvus corax principalis*) in majestic flight. (Author: Copetersen – own work, CC BY-SA 3.0. https://creativecommons.org/licenses/by-sa/3.0/deed.en).

FIGURE 18.4
Group of four common ravens communicating with each other. [Author: Bogomolov PL – own work (Public Domain).]

location of a food source if it stays one night in the roost, strengthening the belief that information is exchanged between ravens nocturnally. Leaders and followers benefit at the food source because new birds may be recruited into the group, and this increases the foraging and food discovery chances. Moreover, large groups might form a better defense against predators and create more mating opportunities. A group of four ravens in North Mongolia communicating with each other is shown in Figure 18.4.

Several species of insects, birds, and animals engage in *social roosting*, where tens, hundreds, or thousands of them come to roost. The roosting time varies from one night to longer periods. These roosting places serve as information centers where information on food sources or about the environment are exchanged. The locations of food sources are not known with certainty among birds and animals and hence they forage for food. This has to be done in an effective manner so that they forage successfully and survive in a competitive environment. Food sources are varying due to changing environmental conditions and depletion (consumption) over time. Therefore foraging has to be adaptive and feedback of past successes has to be incorporated in the search process. Foraging is done either individually or in a collective manner. Social foraging is done collectively by a group of members of the same species. They cooperate with each other to share the food that has been discovered. The scouting strategy, communication of discovered (location) food sources among the members of the group, and sharing of the food discovered are some of the important characteristics of a species. The foraging behavior depends on the location, quality, and quantity of the food source, competition, and risk of predators. Figure 18.5 shows a group of common ravens feeding together on prey.

The main factors involved in social foraging are:

- Searching strategy for food source
- Communication of a discovered food source to other members of the group
- Sharing of the discovered food
- Random component in the movement of the insects/birds/animals

FIGURE 18.5
Group of common ravens feeding. (Author: Henk Monster, CC BY 3.0. https://creativecommons.org/licenses/by/3.0/deed.en.)

Some of the possible ways to communicate discovery of a food source are:

- Broadcast the discovery from the location of food source.
- Generate a visible or invisible trail between the food source and a common location (location of the species, probably at the roost).
- Go to a central location and direct fellow animals or birds or insects to find the food source.
- Go to the central location and recruit fellow animals or birds or insects and lead them back to food source.

The member that discovers the food source and informs others through one of the methods given above has to bear the cost of sharing the food and risks danger in being attacked by a predator. The possible benefits could be collective defense against predators at the food site and attracting of mates. It could also be of benefit if several members are required to attack the prey which could be large in size and dangerous. Some of the species that feed on seeds or fruits stay in groups and hence they forage and advertise the discovery. So the communication of discovery of food sources and sharing of food vary from species to species.

The study of these foraging behaviors of insects/birds/animals has led to the development of various optimization algorithms. The study of foraging behavior in a dynamic environment is the basis of the development of metaheuristic algorithms to solve complex and NP-hard problems efficiently. The *social roosting* behavior of ravens where roosts serve as *information centers* has been applied in developing a novel optimization algorithm, called the *raven roosting* algorithm. This raven roosting algorithm is based on the foraging behavior of ravens where the information about food sources is exchanged or communi-cated to other members of the species in the roost. The *leader* forages and when a quality food resource is found, it leads the other recruited members of the roost to the food source. Raven roosts consist of common ravens, and they come to roost at sunset and leave the roost at dawn the next day. Usually they move in groups. Based on a study undertaken, it is found that when there is large amount of food at a source (huge carcass of an ani-mal), initially a few ravens feed on it. In the following days (assuming it takes a few days to completely eat the carcass) more ravens come to feed on the food source. As the food becomes depleted, the number of ravens at the food source decreases and finally the food is finished. Ravens which were newly added to the roost also leave the roost at dawn and go to the food source to eat. This study has confirmed that exchange of information takes place in the roost after sunset.

18.3 Raven Roosting Optimization

The roosting and social foraging behavior of ravens has led to the development of the raven roosting optimization (RRO) algorithm that is discussed in this section. The forag-ing behavior of ravens and other birds depends on their sensory abilities such as vision, hearing, and smell, remembering the past foraging location (whether successful or not), the ability to transmit or receive information about food source, and randomness in the search for food sources. These characteristics of ravens in foraging have been incorpo-rated into the development of the raven roosting algorithm. In each iteration, a set of birds are recruited by a *leader* which has found the best food source among all the ravens at the roost and they follow the *leader* the next day. Some of the ravens travel towards a food source that was already found by them earlier. If another better food source is discovered on the way, they might change their path to prey upon the newly discovered food source.

18.3.1 Algorithm

A raven roost is located randomly in the search space, and it remains there for the entire run of the algorithm. Assume that there is a population of N ravens and they are randomly located in the search space. Each location of a raven is a potential source of food. The objec-tive (fitness) function defined for the problem is evaluated at each of these locations, and the best solution (fitness value) corresponds to the raven that has the best quality food source and it becomes the *leader* of the population. A group of ravens in the roost among the total population of N ravens are recruited to follow the *leader* to the best food source found by the leader so far (in the previous iteration of the algorithm). This percentage of followers (% of total population N) is represented by P_f. The radius of the sphere around the location of the best food source found by the *leader* is denoted as R_L. The follower birds each forage within this radius at a randomly chosen location, as shown in Figure 18.6.

A group of ravens are newly recruited to the food source discovered by the *leader* whereas the remaining ravens of the roost travel towards an already discovered food source, which may be denoted as their *personal best*. Ignoring the information of discovered food source communicated by the *leader*, they continue to forage at their previously discovered indi-vidual best food destination, as memorized by each one of them. The ravens have memory,

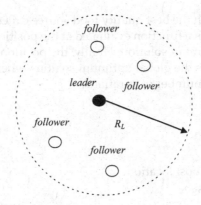

FIGURE 18.6
Sphere around the food source found by the leader.

and they remember their food site where they foraged earlier in a private manner. These birds are neither dependent on the *leader* nor do they follow the *leader*.

The ravens are represented by the *d*-dimensional vector $X_i^{iter} = [x_{i1}^{iter} \; x_{i2}^{iter} \; .... \; x_{id}^{iter}]$, $i = 1, 2, ... N$, where X_i^{iter} is the position of the *ith* raven in the *d*-dimensional search space in iteration represented by the variable *iter*. The flight of ravens could be modeled as a series of finite steps, and the mathematical equation representing the updating of the raven position is given by:

$$X_i^{iter+1} = X_i^{iter} + rand(i).step_i^{iter} \tag{18.1}$$

where X_i^{iter} is the current position of the *ith* raven, X_i^{iter+1} is the next position of *ith* raven, *rand(i)* is a random number with a uniform distribution in the interval [0, 1], and $step_i^{iter}$ is the step size taken by the *ith* raven to reach the new position. The step sizes taken by the ravens during their flight are not uniform since there is a stochasticity included in the movement of the ravens.

Figure 18.7 shows a raven which is following the leader.

During the flight, the *follower* raven forages in a circular region of radius R_f and if a better food source is found, it starts feeding from that location with a probability p_s. Since the food source found is better than that of its *personal best*, the raven updates its *personal best*. The number of perceptions of the raven in the vicinity of the food source is n_p before the raven decides to stop at that location. This process is repeated in multiple iterations, and in

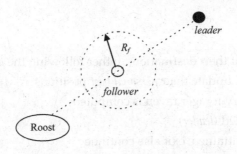

FIGURE 18.7
Recruited raven following the *leader*.

each iteration, the raven with the best quality food source called the *leader* is the best solution found so far. The objective function evaluated at the position of the *leader* in the search space is the *current best* optimum solution. Finally, the position of the raven at the highest quality food source becomes the global optimum solution when the algorithm terminates at the end of the maximum number of iterations.

18.3.2 Pseudocode

Initialization

    Random selection of a roost location

    Population size of ravens N

    Define objective (fitness) function

    Parameters P_f, R_L and R_f, p_s, n_p

    Stopping criteria, if any

    Maximum number of iterations *MaxIter*

    iter = 1

while (*iter* $\leq$ *MaxIter*) **do**

    Population of N ravens are randomly located in the search space

    Objective function is evaluated at each raven location

    personal best location of each raven is updated

    Raven at the location with the highest fitness value is the *leader*

    Percentage P_f of *follower* ravens are recruited to follow the leader and forage
        within a radius R_L of the sphere encircling the leader

    Unrecruited ravens take flight towards their *personal best* locations

    Initialize step size taken by the raven $step^{iter}$

    while ($step^{iter}$ $\leq$ *MaxStep*) **do**

        Every raven that is on the way, either to the *leader's* (best) food source or its own
        personal best, searches in the vicinity of its current position within a radius R_f

        if better food source is found after n_p perceptions

        then raven stops at that location with a probability p_s and updates its *personal best*

        else raven continues its flight using Equation 18.1

        end if

        Increment $step^{iter}$

    end while

    Ravens that arrive at their destinations, either following the *leader* or

    flying on their own, update their *personal best* positions

    Fitness value of every forager raven is computed

    Best fitness is updated (*leader*)

    if stopping criteria attained, exit **else** continue

    iter = *iter* + 1

end while

Raven at the location with highest fitness value is the global optimum solution

Flowchart

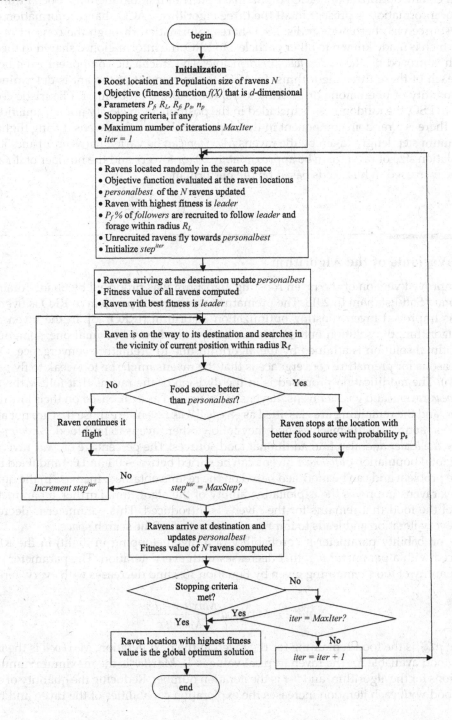

begin

Initialization
- Roost location and Population size of ravens N
- Objective (fitness) function $f(X)$ that is d-dimensional
- Parameters P_f, R_L, R_f, p_s, n_p
- Stopping criteria, if any
- Maximum number of iterations *MaxIter*
- $iter = 1$

- Ravens located randomly in the search space
- Objective function evaluated at the raven locations
- *personalbest* of the N ravens updated
- Raven with highest fitness is *leader*
- $P_f\%$ of *followers* are recruited to follow *leader* and forage within radius R_L
- Unrecruited ravens fly towards *personalbest*
- Initialize $step^{iter}$

- Ravens arriving at the destination update *personalbest*
- Fitness value of all ravens computed
- Raven with best fitness is *leader*

Raven is on the way to its destination and searches in the vicinity of current position within radius R_f

Food source better than *personalbest*?

No → Raven continues it flight

Yes → Raven stops at the location with better food source with probability p_s

$step^{iter}$ = *MaxStep*?

No → Increment $step^{iter}$

Ravens arrive at destination and updates *personalbest*
Fitness value of N ravens computed

Stopping criteria met?

No → $iter$ = *MaxIter*?

Yes → Raven location with highest fitness value is the global optimum solution

No → $iter = iter + 1$

end

Ravens have an in-built capacity for perceiving the environment and sampling locations during flight to find a good food source for foraging and updating their *personal best*. RRO and particle swarm optimization (PSO) share a similarity with respect to the concept of memory. The particles or ravens remember their *personal best* attained so far whereas ants in ant colony optimization (ACO) do not have such individual memory. Social transmission of information is present in all the three algorithms. ACO shares information of the food source via pheromone trails; PSO shares information through the concept of *global best* which is made known to other particles. In RRO, the information is shared in the roost which is proved by the *leader–follower* principle. The stochastic component is embedded into each of these three algorithms. In ACO, the movement of the ant is determined by the quantity of pheromone or information specific to the problem or a heuristic component. In PSO, the randomness is included in the position and velocity update equations. In RRO, there is a random component in the step size taken by the ravens during flight. The maximum step length taken by the ravens *MaxStep* can be varied between 10 and 50. The population size of ravens can be approximately 50 (or lower), and the number of iterations can be increased in hundreds based on the performance.

18.4 Variants of the Algorithm

An improved version of the raven roosting algorithm was proposed by Shadi Torabi and Faramarz Safi-Esfahani in 2018. The premature convergence problem in RRO is overcome in this improved raven roosting optimization algorithm (IRRO) [5]. In the raven roosting algorithm, the solution obtained is not always the best or optimal one; sometimes a suboptimal solution is attained by the algorithm due to premature convergence. One of the reasons for premature convergence is that the ravens might be too weak to fly (global search). The modification proposed is in the division of the ravens that follow the *leader* into *weak* ravens and *greedy* ravens. Ravens are classified as *weak* based on their low fitness values, and the remaining are classified as *greedy*. *Weak* ravens are those that are not able to find food sources by themselves, and they follow other ravens to find food. *Greedy* ravens follow the *leader* and also find additional food sources. The percentage of *weak* ravens out of the total population of *follower* ravens can be varied between 0.1 and 0.9 and fixed based on the problem and can be modified based on the results obtained. Increase in the number of *weak* ravens improves the exploitation ability of the algorithm. Further, a parameter to control the food that remains for the ravens is introduced. This parameter is decreased with every iteration and leads to improved exploration of the search space.

The probability parameter p_s (probability of a raven stopping in flight) in the RRO is replaced with a parameter p_{food} that decreases with every iteration. The parameter p_{food} is the quantity of food remaining given by Equation 18.2 and decreases with every iteration.

$$p_{food}^{iter+1} = MaxFood \frac{MaxIter - iter}{MaxIter} \qquad (18.2)$$

where p_{food}^{iter+1} is the food remaining for the raven in the next iteration, *MaxFood* is the maximum food available for the raven (typical value is 1), *MaxIter* is the maximum number of iterations for the algorithm, and *iter* is the iteration number. Reducing the quantity of available food with each iteration increases the exploration capabilities of the raven and hence the diversification of the algorithm. The IRRO algorithm has been found to be better than or comparable to RRO, PSO, chicken swarm optimization (CSO), bat algorithm (BA), gray wolf optimization (GWO), and whale optimization algorithms (WOA).

18.5 Summary

Metaheuristic optimization algorithms provide solutions to complex problems in finite time. The optimal solutions are usually selected from a set of possible as well as feasible solutions to the problem. The importance of metaheuristics is that it arrives at the optimum in practically reasonable time although the optimum solution might not be an exact solution to the problem. The global optimum solution produced by the algorithm could be quite close to the best and accurate solution to the problem. Most of the metaheuristic algorithms have reasonable time and computational complexity. The optimization algorithm should have balance between intensification and diversification. The RRO algorithm discussed in this chapter is based on the social roosting and foraging behavior of common ravens. The RRO algorithm converges prematurely since it does not have good diversification property. Moreover, sensitivity analysis shall give an insight into the effect on the various components of the RRO algorithm. It is found that RRO is quite competitive with other similar algorithms. The premature convergence problem in RRO has been overcome in the improved RRO algorithm. The IRRO algorithm improves the global exploration capability of the RRO algorithm, by introducing the food parameter and the concept of weak ravens that takes care of the exploitation capabilities.

References

1. Anthony Brabazon, Wei Cui, Michael O'Neill, The raven roosting optimisation algorithm, *Soft Computing* (Springer-Verlag), Vol. 20, pp. 525–545, 2016.
2. J. Wright, R. Stone, N. Brown, Communal roosts as structured information centres in the raven, *Corvus Corax*, *Journal of Animal Ecology*, Vol. 72, No. 6, pp. 1003–1014, 2003.
3. Richard B. Stiehl, Observations of a large roost of common ravens, *The Cooper Ornithological Society, Condor*, Vol. 83, pp. 78, 1981.
4. John M. Marzluff, Bernd Heinrich, Raven roosts are still information centres, *Animal Behaviour*, Vol. 61, pp. F14–F15, 2001.
5. Shadi Torabi, Faramarz Safi-Esfahani, Improved raven roosting optimization algorithm, *Swarm and Evolutionary Computation* (Elsevier), Vol. 40, pp. 144–154, 2018.

19

Applications

19.1 Introduction

The various optimization algorithms can be compared based on their performance. The algorithms have to be tested and their performance evaluated against some defined standard metrics. The performance measures for testing the algorithms are convergence rate, accuracy of the optimum solution, number of iterations, population size, diversity, and ability to jump out of local optima. These metrics can be quantified in terms of time complexity, space complexity, and computational complexity. Standard benchmark test functions are used for testing the performance of the algorithms and comparing against each other. The standard set of functions in the literature have been taken from various CEC benchmark data sets. The typical engineering design problems such as three bar truss and welded beam design, and computer science problems such as the traveling salesman problem and graph coloring have been given that can be used for testing the algorithms.

19.2 Benchmark Test Functions

The standard test functions applied for testing, validating, and comparing the performance of the different optimization algorithms are enumerated in this section. The functions listed here are the standard functions available in the literature on optimization algorithms, especially for testing the swarm intelligence algorithms. Other than the functions described here, a few other test functions have been used and are available in the literature. The test functions are a mix of unimodal as well as multimodal functions.

De Jong's f1 function is a sphere function that is unimodal and convex and is the simplest in the De Jong's test set. It is defined as

$$f_1(X) = \sum_{i=1}^{d} x_i^2$$

where $X = \{x_1 \; x_2 \; ... \; x_d\}$. The function has a global minimum at $X^* = \{0, 0, ..., 0\}$ with $f(X^*) = 0$.

De Jong's f2 function

$$f_2(X) = 100(x_1^2 - x_2)^2 + (1 - x_1)^2$$

De Jong's f4 function

$$f_4(X) = \sum_{i=1}^{d} i x_i^4$$

Schaffer's f6 function

$$f_6(X) = 0.5 + \frac{\left(\sin\sqrt{x^2 + y^2}\right) - 0.5}{\left(1.0 + 0.001\left(x^2 + y^2\right)\right)^2}$$

Ackley function

$$f(X) = -20\exp\left[-\frac{1}{5}\sqrt{\frac{1}{d}\sum_{i=1}^{d} x_i^2}\right] - \exp\left[\frac{1}{d}\sum_{i=1}^{d}\cos(2\pi x_i)\right] + 20 + e$$

The function is multimodal and has a global minimum at $X^* = \{0, 0, ..., 0\}$ with $f(X^*) = 0$.

Easom's function

$$f(X) = -\cos(x)\cos(y)\exp\left[-(x-\pi)^2 + (y-\pi)^2\right]$$

This function has global minimum $f(X^*) = -1$ at $X^* = (\pi, \pi)$ within the range $-100 \le x,\ y \le 100$. This function is multimodal with several local minima.

Griewangk function

$$f(X) = \frac{1}{4000}\sum_{i=1}^{d} x_i^2 - \prod_{i=1}^{d}\cos\left(\frac{x_i}{\sqrt{i}}\right) + 1 \qquad -600 \le x_i \le 600$$

The function has a global minimum at $X^* = \{0, 0, ..., 0\}$ with $f(X^*) = 0$. This is a multimodal function.

Michaelwicz's function

$$f(X) = -\sum_{i=1}^{d}\sin(x_i)\left[\sin\left(\frac{i x_i^2}{\pi}\right)\right]^{2m}$$

where $m = 10, 0 \le x_i \le \pi$, *for* $i = 1, 2, ...d$.
   The equivalent 2D function is

$$f(x, y) = -\sin(x)\sin^{20}\left(\frac{x^2}{\pi}\right) - \sin(y)\sin^{20}\left(\frac{2y^2}{\pi}\right)$$

where $(x, y) \in [0,5] \times [0,5]$. This function has a global minimum $f(X^*) \approx -1.8013$ at $(x_*, y_*) = (2.20319, 1.57049)$.

Rastrigin's function is a multimodal function defined as

$$f(X) = 10d + \sum_{i=1}^{d} \left[x_i^2 - 10\cos(2\pi x_i) \right] \cdots -5.12 \leq x_i \leq 5.12$$

The function has a global minimum at $X^* = \{0, 0, ..., 0\}$ with $f(X^*) = 0$.

Rosenbrock's function

$$f(X) = \sum_{i=1}^{d-1} \left[(x_i - 1)^2 + 100 \left(x_{i+1} - x_i^2 \right)^2 \right]$$

whose global minimum $f(X^*) = 0$ occurs at $X^* = (1, 1, ..., 1)$ in the domain $-5 \leq x_i \leq 5$ where $i = 1, 2, ..., d$.

The 2D *Rosenbrock's function* is given by

$$f(x, y) = (x-1)^2 + 100 \left(y - x^2 \right)^2$$

This is referred to as a *banana* function.

Schwefel's function

$$f(X) = -\sum_{i=1}^{d} x_i \sin\left(\sqrt{|x_i|} \right) \quad -500 \leq x_i \leq 500$$

whose global minimum at $f(X^*) \approx -418.9829d$ occurs at $x_i = 420.9687$ where $i = 1, 2, ..., d$.

Yang's function

$$f(X) = \left(\sum_{i=1}^{d} |x_i| \right) \exp\left(-\sum_{i=1}^{d} \sin\left(x_i^2 \right) \right) \cdots -2\pi \leq x_i \leq 2\pi$$

This is a forest function with a global minimum $f(X^*) = 0$ at $[0, 0, ..., 0]$.

Shubert's function

$$f(X) = \left[\sum_{i=1}^{d} i\cos\left(i + (i+1)x \right) \right]\left[\sum_{i=1}^{d} i\cos\left(i + (i+1)y \right) \right]$$

The function has many global minima with $f(X^*) = -186.7309$ for $d = 5$ in the domain $-10 \leq x, y \leq 10$.

19.3 Applications

The standard engineering design problems and applications in computer science that are intractable to the classical algorithms are solved easily using nature-inspired swarm-based algorithms. Some of the important applications in the category are given below.

19.3.1 Traveling Salesman Problem

Given a set of cities and distance between each pair of cities, the problem is to find the shortest route in visiting all the cities once without retracing and finally reaching the starting city. This is called a Hamiltonian Tour.

The cities are represented as nodes, and the connecting routes are the edges between nodes with a cost associated with every edge. This set of nodes and edges forms a connected graph. The initial starting place is randomly chosen from the set of cities and the traveling salesman moves from one city to the next with the cost being cumulative. This is a standard combinatorial optimization problem that has been found to be NP-hard for the traditional algorithms to solve. TSP is one of the important problems in optimization and operations research and serves as a benchmark for comparing the performance of the various algorithms.

19.3.2 Knapsack Problem

Given a set of items each with a weight and value, the problem is to fill a rucksack or knapsack such that the total value of the items is maximized with the constraint that the total weight of all the items in the knapsack should not exceed a maximum of W kg.

The number of items could be an integer or a fractional number such as $2x$ (two numbers of the item x) or $0.5x$ (half of the item x) depending on the application. This standard knapsack problem can be applied in real-life situations for solving problems related to decision making, and the algorithm normally used to solve this problem is the greedy algorithm. There is no polynomial time algorithm for solving this problem that has been found to be NP-hard.

19.3.3 Graph Coloring Problem

Given a graph and a set of n number of colors, the problem is to color the vertices of the graph such that no two adjacent vertices have the same color. A variation of this graph coloring problem is to color the edges of the graph instead of the vertices such that no two adjacent edges have the same color.

The number of colors depends on the nature of the problem. It is popularly known as the problem *soduku*, and it is one of the challenging applications in computer science.

19.3.4 Job Scheduling Problem

Given a set of n jobs to be processed and m number of machines, the problem is to schedule the jobs on the various machines so that the total processing time is minimized. Every job might have a finite number of operations to be carried out in time sequence, and each of the machines might have different processing power to carry out the operations.

This is one of the standard combinatorial optimization problems in computer science that has been found to be NP-hard.

19.3.5 Feature Reduction Problem

Given a set of features or attributes of an object, the problem is to select the minimum number of features such that the accuracy of the object classification is maximized.

This is one of the important dimensionality reduction problems in machine learning and image classification where the dimension of the search space is reduced, thus reducing the time and computational complexity. Several transformation techniques are available that remove the redundancy in the data and reduce the number of components to be processed. If an exhaustive search of the feature space is to be conducted it will take up lot of time.

19.3.6 Network Routing Problem

Given a computer or communication network, the problem is to find an optimal route through the network such that the total cost is minimized under constraints, if any. The constraints could be that every node of the network is to be visited once or it could be with respect to distance.

This is one of the earliest problems in computer science that has received a lot of attention in the field of optimization and operations research. The problem of routing packets in computer networks has been a fertile research area for several years, and it continues to be so. Given a choice of links in a network which could even be a telecommunication network, the problem is to identify the best route in terms of minimizing total cost incurred.

19.4 Summary

The different nature-inspired algorithms can be tested and compared with performance metrics that are defined for the optimization algorithm. Each algorithm is suitable for a particular class of applications, and the performance might not be equally good for another set of problems. This is the No Free Lunch theorem that has been discussed earlier. The MATLAB codes for some of these algorithms can be downloaded from the *MathWorks* website, www.mathworks.com, to test their performance for various real-time applications.

20

Conclusion

An introduction to the general theory of optimization and its mathematical formulation has been discussed in Chapter 1. An overview of classical and traditional optimization algorithms has been given in Chapter 2. The third chapter deals with the nature-inspired optimization algorithms, their characteristics, and their advantages and disadvantages. The different nature-inspired, swarm intelligence algorithms have been described in Chapters 4 to 18, and some of the popular applications have been listed in Chapter 19. Moreover, the standard benchmark data sets that can be used for testing and validating the algorithms are also given in Chapter 19. The book concludes with a summary in Chapter 20.

The traditional optimization methods have been discussed in Chapter 2, and a few examples have been given wherever possible. The popular traditional algorithms such as linear programming, non-linear programming, quadratic programming, geometric programming, dynamic programming, integer programming, and stochastic programming have been outlined. The various algorithms are classified based on the number of objective functions, number and type of constraints, number of variables involved, characteristics of the objective function, and so on. The optimization method to be applied depends on the problem and its mathematical formulation, and most of them are numerical programming methods. The choice of the algorithm depends to a great extent on the dimension of the problem to be solved. The classical engineering design problems such as design of steel and civil structures, optimizing input resources or maximizing output from manufacturing industries, food-processing industries, and chemical industries, and routing in computer and communication networks are some of the diverse applications of the traditional optimization techniques.

Nature-inspired algorithms are metaheuristic and are found to be able to solve challenging problems efficiently. Many of the real-world applications are highly non-linear, requiring state-of-the-art optimization techniques for solving them. Nature-inspired algorithms are flexible, adaptive, self-organized, and population-based with simple interactions among search agents. Metaheuristic algorithms outperform their traditional counterparts due to their search in parallel by a population of agents, the lower number of parameters to tune, the ease and simplicity of implementation, and the dynamic shift between exploration and exploitation phases. Randomization by following some stochastic distribution such as uniform or Gaussian helps in achieving diversity of solutions. Increasing the diversity of solutions reduces the possibility of getting trapped in local optima, and hence these nature-inspired algorithms can deal with complex problems very efficiently with least time complexity.

The biological evolution principles in optimization algorithms have led to the development of GA, whose performance has been validated for a myriad of complex problems over the years since its inception. GA uses an objective function but does not require the computation of derivatives of the objective function. It uses stochastic rules while searching, by introducing controlled randomness into the algorithm instead of being completely deterministic as in the classical optimization algorithms. GA tries to mimic the human evolution process and searches for an optimum solution with a population of individuals.

This implicit parallelism in GA helps in searching and exploiting large regions of the search space simultaneously. Typical parameters in GA could be population size of 40, maximum number of generations 20, and mutation rate less than 0.1 with the remaining (0.9) being crossover. The encoding and length of the chromosome depend on the problem, and randomness is introduced in the selection, crossover, and mutation operations.

Genetic programming is a variant of GA that incorporates biological evolution and computer science. GP involves populations of computer programs that evolve which is an advanced technique compared to other evolutionary algorithms. Typical population size could be less than 1000 for small problems, but it can increase beyond that for problems of large size. The maximum number of generations can be initially 20 and could go up to 100 or even beyond that if the problem requires so. The inherent parallelism in GP and its rate of convergence play an important role in the applications of GP, and it can accommodate data-intensive applications.

In PSO the population remains constant throughout the run of the algorithm unlike GA where the population changes with every generation. The typical population size is 10 to 50, chosen depending on the problem. The uniqueness of PSO lies in the fact that particles fly through the solution space accelerating towards better solutions. As the number of dimensions increases, the algorithm converges slowly. The parameters c_1, c_2, w_v may be held constant for all the iterations, or c_1, c_2 may be held constant with linearly decreasing inertia weight w_v. An initial large inertia weight leads to exploration of the search space, and as the weight decreases, it increases the exploitation abilities of the swarm. There are only a few parameters to be controlled, and it is computationally efficient, derivative-free, simple to implement, and applicable to a wide range of problems.

DE is a global optimization technique that is easy to apply, reliable, fast, and simple to implement. DE can efficiently handle unimodal as well as multimodal objective functions that are non-linear and non-differentiable. DE is robust with few parameters, it converges consistently, and it is faster and computationally more efficient than other classical optimization methods. Differential evolution was proposed to be a stochastic direct-search method to find the global minimum, with inherent parallelism to take care of computationally intensive cost functions. DE has a self-organizing capability that makes it remarkably different from other optimization techniques, and its implicit parallelism increases the rate of convergence. DE has been proved to be efficient in solving engineering design problems and operates on continuous spaces.

Inspiration from the swarm intelligence of ants has led to the development of the ACO algorithm. The collective intelligence of a swarm of ants has been used to solve problems like graph-based searches, where finding the shortest path through an interconnected graph is an NP-hard problem. ACO has the ability to rapidly converge on the optimum solution for discrete combinatorial problems. When the set of decision variables is large, the ACO algorithm will be able to find the optimum solution in finite time. ACO is suitable for problems like TSP and other NP-hard problems for which the dimension of the problem is increasing exponentially. Typical problems that require finding the shortest path through a network can be solved efficiently with ACO rather than other greedy algorithms.

The BCO algorithm is capable of solving difficult combinatorial optimization problems. It is a metaheuristic algorithm that has been inspired by the behavior of honey bees in nature. Bees exhibit collective intelligence in collecting nectar and producing honey, and study of their behavior and activities has motivated the development of a set of optimization algorithms to solve complex engineering design problems. Since the first algorithm was proposed in 1997 several variants have been developed and applied to different problems in engineering and computer science. The performance of the algorithm shows that simple insects like bees can inspire and motivate us to develop algorithms to solve complex problems. The individual and collective complex behaviors of honey bees have been mimicked in developing these algorithms. The optimization algorithm based on the behavior of honey bees has been successfully applied to several real-life problems such as job shop scheduling, clustering, image analysis, optimal design of structures, and complex engineering problems.

FSS is a novel metaheuristic search algorithm based on the swarming behavior of fish that is able to solve complex problems with simple mathematical models. The FSS algorithm also has its own specific characteristics and properties that make it suitable for high-dimensional, unstructured, multimodal search spaces. The fish schools contain high volume and density in some of the species and could go up to thousands of fish. On average, a population size of 20 to 50 with a maximum of 100 iterations is sufficient for solving most problems. FSS has a good balance between exploration and exploitation abilities and produces excellent results for unimodal as well as multimodal NP-hard problems. FSS has been experimentally found to outperform PSO for standard benchmark data sets.

The cuckoo search algorithm is based on the parasitic breeding behavior of cuckoo birds combined with Levy flights. The population size N and the probability p_h are two parameters of the cuckoo search algorithm that need to be chosen for the problem, and it has been found that $N = 15$ and $p_h = 0.25$ are sufficient for most of the problems. As the algorithm runs, the nests aggregate at the global optimum for unimodal functions, and when the function is multimodal, the nests distribute themselves at positions of the local optima. The convergence rate of the algorithm does not depend on the above parameters and makes it suitable for NP-hard, single- and multi-objective problems. Compared to PSO and GA, the cuckoo search algorithm outperforms them for all benchmark unimodal and multimodal test functions. It is superior because it allows a diversity of solutions and is also able to intensify the search in local regions by means of Levy flights. The cuckoo search algorithm hybridized with other nature-inspired algorithms and its variants have been found to be more powerful in solving tough optimization problems.

FA is a swarm intelligence optimization algorithm based on the swarming behavior of fireflies and exhibits the characteristics and properties of other swarm intelligence algorithms. FA can efficiently solve continuous as well as discrete combinatorial optimization problems. In FA multiple fireflies search the space in parallel, and this inherent parallelism improves the efficiency of the algorithm. The firefly algorithm is more efficient in dealing with multimodal optimization problems. Initially, the number of fireflies N should be distributed uniformly over the entire search space, and, within 50 to 100 iterations, convergence takes place. The typical population size used in most of the applications is $N = 20$ to 40. The greater the number of fireflies, the faster will be the convergence. For most of the problems, experimentally it has been found that the parameters can be chosen as c in the range of $[0, 1]$, $A_0 = 1$, $\lambda = 1$, $u = 1.5$. The change in attractiveness of the firefly with distance is characterized by the parameter λ, and it determines the speed of convergence of the algorithm. The attractiveness parameter λ typically varies from 0.01 to 100. The firefly algorithm with Levy flights has been found to outperform GA and PSO, and PSO is a special case of the FA for certain settings of the parameters. It has reduced time and computational complexity with few parameters to be tuned.

The bat algorithm is one of the superior nature-inspired optimization techniques that uses the echolocation principle and Doppler effect to detect and locate prey. The associated parameters of frequency of pulses emitted, rate of emission, and loudness of emission can be adjusted and fine-tuned to vary the performance. The right combination of these parameters plays a key role in finding the global optimum solution to the problem. By adjusting the average loudness and rate of emission the BA algorithm effectively reduces to either PSO or Harmony Search. A population size of 20 to 50 and 100 iterations is sufficient for most of the applications. BA is simple, efficient, easy to implement, and flexible and can be applied to wide range of problems. The parameters can be fine-tuned or modified adaptively as the bat is approaching the prey, that is, as the algorithm is nearing the optimal solution or as the iterations proceed it automatically changes from the exploration to exploitation mode. This makes BA more efficient than other algorithms. Bats have a signal-processing capability that is different from other animals and insects, and BA is more powerful than GA and PSO because it includes the important characteristics and advantages of these algorithms as well as of other nature-inspired algorithms such as Harmony Search and Simulated Annealing.

The flower pollination algorithm is based on the pollination process of plants or flowers. FPA is also population-based and searches for the optimum solution in parallel in the search space. The algorithm can be employed for unconstrained as well as constrained optimization problems. The algorithm is simple with few parameters to tune, and it can be applied for any complex engineering design problem. The number of parameters in FPA is less, and its performance is comparable to or surpasses other metaheuristic algorithms such as GA and PSO. This simplicity makes it very popular for solving NP-hard optimization problems. A population size of around 20 to 50 with 100 iterations is suitable for most applications. The simple concept of one pollen gamete and one flower on one plant can be extended to multiple pollen gametes and multiple flowers depending on the applications.

The gray wolf optimization algorithm is a metaheuristic optimization algorithm that emulates the hierarchy of the wolf pack and their hunting techniques employed. The hunting, chasing, encircling, and attacking of prey by gray wolves are the fundamental operations employed in the GWO algorithm. GWO has been applied on several benchmark data sets and engineering design applications and found to outperform the existing algorithms for solving complex problems. GWO has been found to be simple, flexible, derivative-free, and efficient in attaining the global optimum solution. The performance of the GWO algorithm is competitive with particle swarm optimization, differential evolution, and the gradient search algorithm and even outperforms some of them for a few benchmark functions. The GWO algorithm is found to have good exploration, exploitation, avoidance of local optima, and convergence to the optimum solution and is suitable for real applications.

The intelligence and smart behavior exhibited by elephants have been the source of inspiration for the development of the EHO algorithm. It has been successful in providing solutions to complex engineering problems that are quite challenging and time-consuming to solve by other traditional methods. EHO has been tested on various benchmark problems and found to be competitive with the existing optimization algorithms such as GA and DE. Typical values for the parameters of the EHO algorithm are $N_C = 5$, $N_{CE} = 10$, and this makes the total elephant population size 50. The scale factor for influence of the matriarch on the clan updating operation is $s_m = 0.5$, and the factor $r = 0.1$ is the influential parameter for matriarch updating. The dimension of the population space could be between 10 and 20, and the maximum number of iterations could range from 50 to 100, depending on the problem to be solved. The EHO algorithm has fewer parameters to control compared to other swarm-based algorithms. The EHO algorithm is easy to implement since it has a lower number of parameters compared to other optimization algorithms. EHO gives a good performance on most of the benchmark test functions compared to GA and DE.

Crows are very intelligent birds that exhibit swarm behavior, and they have been the source of inspiration behind the crow search optimization algorithm. The performance of the algorithm depends on two important parameters – flight length and awareness probability. In addition, some randomness is introduced in the algorithm by the two random parameters r_i and r_k that follow uniform distribution. The awareness probability is an important parameter associated with CSA that provides the necessary balance between intensification and diversification. The CSA and its proposed variants applied to solve various engineering optimization problems that are available in the literature have also been outlined. This simple algorithm has been found to be effective in solving optimization problems that are NP-hard in terms of convergence speed and computational complexity. CSA has been found to give good performance when compared to GA and PSO. CSA is robust and optimum results can be mostly obtained in less than 50 iterations on average. There are only two main parameters to be tuned for the algorithm which is an added advantage. Typical values for population size are between 20 and 40, awareness probability is <0.2, and flight length is <2. CSA is simple and easy to implement and gives good performance over a wide spectrum of complex optimization problems.

The social roosting and foraging behavior of the common raven is the inspiration behind the development of the raven roosting optimization algorithm. The RRO algorithm converges prematurely since it does not have good diversification properties, and this drawback has been overcome in the improved RRO algorithm. It is found that RRO is quite competitive with other similar algorithms. The IRRO algorithm improves the global exploration capability of the RRO algorithm, by introducing the food parameter and the concept of weak ravens that takes care of the exploitation capabilities. Every nature-inspired algorithm has its own specialized features that are suited to a specific set of problems.

The standard benchmark functions used for testing and validating the performance of the various optimization algorithms have been listed in Chapter 19. These are the commonly used functions for comparing the performance of the nature-inspired optimization algorithms. The important functions that are a mix of unimodal and multimodal objective functions are given. The typical important applications that are NP-hard and found to be intractable for the classical and traditional algorithms to solve are also described. These are the applications which the swarm intelligence algorithms are designed to solve with least time and computational complexity. The superiority of the nature-inspired algorithms has been established by quantifying their performance on these benchmark test functions and standard applications. The algorithms can be implemented in MATLAB and for a few of these, the MATLAB code can be downloaded from www.mathworks.com.

Index

Printed in the United States
by Baker & Taylor Publisher Services